Wearable and Neuronic Antennas for Medical and Wireless Applications

Scrivener Publishing
100 Cummings Center, Suite 541J
Beverly, MA 01915-6106

Advances in Antenna, Microwave, and Communication Engineering

Series Editor: Manoj Gupta, PhD, Pradeep Kumar, PhD

Scope: Antenna and microwave, as well as digital communication, engineering has been increasingly adopted in many diverse applications such as radio astronomy, long-distance communications, space navigation, radar systems, medical equipment's as well as missile electronic systems. As a result of the accelerating rate of growth of communication, microwave and antenna technology in research and industry sectors; students, teachers and practicing engineers in these area are faced with the need to understand various theoretical and experimental aspects of design and analysis of microwave circuits, antennas and simulation techniques, communication systems as well as their applications. Antennas, Microwave and Communication Engineering are actually a very lively and multidisciplinary one, mixing the deepest electromagnetic theoretical aspects, mathematical signal and data processing methods, physics of devices and physics of fields, software developments and technological fabrication aspects, and the large number of possible applications generates multiple outcomes. Hence the aim of this book series is to provide a multi-discipline forum for engineers and scientists, students, researchers, industry professionals in the fields of Antenna, Microwave, Communication and Electromagnetic Engineering to focus on advances and applications.

Publishers at Scrivener
Martin Scrivener (martin@scrivenerpublishing.com)
Phillip Carmical (pcarmical@scrivenerpublishing.com)

Wearable and Neuronic Antennas for Medical and Wireless Applications

Edited by

Arun Kumar
Manoj Gupta
Mahmoud A. Albreem
Dac-Binh Ha

and

Mohit Kumar Sharma

Scrivener
Publishing

WILEY

This edition first published 2022 by John Wiley & Sons, Inc., 111 River Street, Hoboken, NJ 07030, USA and Scrivener Publishing LLC, 100 Cummings Center, Suite 541J, Beverly, MA 01915, USA
© 2022 Scrivener Publishing LLC
For more information about Scrivener publications please visit www.scrivenerpublishing.com.

Wiley Global Headquarters

111 River Street, Hoboken, NJ 07030, USA

For details of our global editorial offices, customer services, and more information about Wiley products visit us at www.wiley.com.

Limit of Liability/Disclaimer of Warranty

Library of Congress Cataloging-in-Publication Data

ISBN 978-1-119-79180-5

Cover image: Pixabay.com
Cover design by Russell Richardson

Set in size of 11pt and Minion Pro by Manila Typesetting Company, Makati, Philippines

Printed in the USA

10 9 8 7 6 5 4 3 2 1

Contents

Preface

Wearable antennas, neuronic antennas, medical antennas, microstrip, Jeans substrate, wireless body area network, on-body transmission antennas, body phantom, dual-mode antennas, on-body path loss, wearable microstrip antennas, antennas for medical applications, body-centric wireless communications antennas, robust planar textile antennas, dual-mode antennas for on off-body communications, modular textile antennas, textile patch antennas, E-textile conductors and polymer composites, inductively coupled fed loop antennas, multiple inputs and multiple outputs (MIMO) 5G antennas, WLAN antennas, Internet of Things (IoT) antennas, wire antennas, and aperture antennas will play an important role in wireless applications. The signals created by human cerebrum neurons can be enhanced a few thousand times by methods for a gadget currently known as the MIND SWITCH. The term "brain change" refers to the technology that has been developed that allows an individual to kill on an electrical apparatus, for example, a work area light or TV, in 2–3 seconds using EEG signals without any preparation. Relative control, such as increasing or decreasing the volume of a proportion, is also possible with the technology. The book begins with an overview of advanced waveforms. The features of the 5G waveform and antenna design are then defined, comprising their major capability restrictions. Various designs of antenna, waveform coding, and signal processing structures are then deliberated in the feature, containing state-of-the-art neuronic antenna, medical antenna, micro strip, Jeans substrate, wireless body area network, on-body transmission antenna, body phantom, dual-mode antenna, on-body path loss, wearable microstrip antenna, antennas for medical applications, body-centric wireless communications antenna, and multiple antenna techniques. The final chapters described the filter designs and wearable antennas. Design comprehension and tradeoffs are highlighted throughout the book. It covers numerous worked examples, over 300 figures and over 800 references, and is a perfect textbook for researchers, academicians, and students.

Machine Learning Aided Channel Equalization in Filter Bank Multi-Carrier Communications for 5G

Ubaid M. Al-Saggaf[1,2], Muhammad Moinuddin[1,2]*, Syed Saad Azhar Ali[3], Syed Sajjad Hussain Rizvi[4] and Muhammad Faisal[5]

[1]Center of Excellence in Intelligent Engineering Systems (CEIES), King Abdulaziz University, Jeddah, Saudi Arabia
[2]Electrical and Computer Engineering Department, King Abdulaziz University, Jeddah, Saudi Arabia
[3]Center for Intelligent Signal and Imaging Research (CISIR), Electrical and Electronics Engineering Department, Universiti Teknologi PETRONAS, Bandar Seri Iskandar, Malaysia
[4]Computer Science Department, SZABIST, Karachi, Pakistan
[5]Computer & Information Technology Dept., Dammam Community College, King Fahd University of Petroleum & Minerals, Dhahran, Saudi Arabia

Abstract

Multi-carrier communications (MC) have gained a lot of interest as they have shown better spectral efficiency and provide flexible operation. Thus, the MC are strong candidates for the fifth generation of mobile communications. The Cyclic-prefix orthogonal frequency division multiplexing (CP-OFDM) is the most famous technique in the MC as it is easy to implement. However, the OFDM has poor spectral efficiency due to limited filtering options available. Thus, to enhance spectral efficiency, an alternative to OFDM called Filter bank multicarrier (FBMC) communication was introduced, which has more freedom of filtering options. On the other hand, the FBMC preserves only real orthogonality for the waveforms, resulting in imaginary interference. Hence, the equalization in FBMC has to deal with this additional interference which becomes challenging in multiuser communication. In this chapter, the aim is to deal with this challenge.

*Corresponding author: mmsansari@kau.edu.sa

Arun Kumar, Manoj Gupta, Mahmoud A. Albreem, Dac-Binh Ha and Mohit Kumar Sharma (eds.)
Wearable and Neuronic Antennas for Medical and Wireless Applications, (1–10) © 2022 Scrivener Publishing LLC

Keywords: Multiuser communications, multicarrier communications, OFDM, FBMC, 5G, equalization, machine learning, MMSE

1.1   Introduction

Improved bandwidth, efficient power utilization, and better handling of selective fading are prominent advantages of the MC [1]. The CP-OFDM is a simpler MC solution. However, it has poor spectral efficiency due to limited filtering options available.

Another candidate considered for multiple radio access in 5G is the Filter Bank Multi-Carrier (FBMC) [2]. The FBMC is a modified version of OFDM, where the rectangular transmit and receive waveforms are replaced by any other more frequency localized pulse shapes, which offer a more confined spectrum, relevant for spectrum-sharing scenarios.

In the FBMC with Offset Quadrature Amplitude Modulation (OQAM) system, Channel equalization is challenging compared to the standard CP-OFDM. This is because the real and the imaginary symbols in the FBMC-OQAM system are transmitted with a time offset, which results in loss of orthogonality in the imaginary part [2]. Thus, the received signal suffers from the additional imaginary part, which is termed as Imaginary interference. This interference degrades the performance of the channel estimator. As a result, channel equalization also degrades as it requires the knowledge of channel estimates. In conventional equalization methods for the FBMC-OQAM, the channel estimates are used to cancel the phase of the received signal and extract its real part. However, due to errors in channel estimates, the phase cancellation is not completely achieved. Thus, this conventional approach does not work in this scenario. Alternatively, people have proposed direct equalization methods, which still need improvements. Therefore, there is a need to design an efficient equalization method to combat the imaginary interference and inter-symbol interference (ISI), and inter-carrier interference (ICI) in the FBMC-OQAM system.

1.2   Related Literature Review

The FBMC modulation is a recent type of MC technique developed to improve the performance of the conventional OFDM by minimizing ISI and ICI [1, 2].

The FBMC-OQAM has been considered a strong candidate for 5G due to its better spectral efficiency in terms of out-of-band (OOB) emission in contrast to the conventional OFDM. On the other hand, the FBMC-OQAM loses orthogonality in the imaginary part due to offset transmission scheme, resulting in imaginary interference [2]. Thus, more sophisticated methods are required to deal with channel estimation [3], equalization, and interference cancelation in Multiple-Input and Multiple-Output (MIMO) FBMC implementation [4–6].

The simplest equalization method for the FBMC system is the one-tap equalizer which has shown better performance in the absence of ISI and ICI [7]. However, its performance severely degrades in the presence of selective channels, particularly for high Signal-to-Noise Ratio (SNR) scenarios.

The existing solutions for equalization in the FBMC system commonly assumed that the channel is time-invariant [8–12]. The equalization method in [8] developed a parallel architecture using Fast Fourier Transform (FFT) blocks for the equalization. Another method based on larger FFT was proposed in [10]. In [11], a Minimum Mean Squared Error (MMSE) criteria-based equalization technique was developed, which was later modified in [12] to the MIMO scenario. In [13], a spatio-temporal structure based n-tap MMSE equalizer was introduced for the FBMC systems.

The authors in [14] have given evidence that the MMSE equalizer for the FBMC has robustness property in the presence of a time-varying channel.

In [15], the authors have reformulated the multicarrier equalization problem as single carrier QAM and derived relevant MMSE expression.

1.3   System Model

In the FBMC-OQAM transmission, various processes are involved, as shown in Figure 1.1.

It can be seen from Figure 1.1 that the FBMC-OQAM transmitter consists of a first block of bits to symbol mapping, which is followed by a OQAM processing block that modulates the symbols using OQAM mapper. Next, the IFFT is taken after converting data from serial to parallel. Finally, the data is transmitted through the channel by converting it again

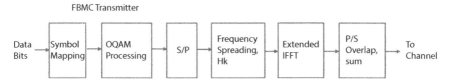

Figure 1.1 FBMC-OQAM transmission system.

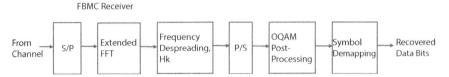

Figure 1.2 FBMC-OQAM reception system.

to serial transmission. The FBMC receiver has reverse processing of transmission blocks, as shown in Figure 1.2.

The transmitted basis pulse for the lth sub-carrier and kth symbol $g_{l,k}(t)$ is defined as:

$$g_{l,k}(t) = p(t - kT)\frac{j2\pi}{e^{F(t-kT)}e^{j\theta l.k}} \tag{1.1}$$

Where $p(t)$ is the prototype filter, F is the subcarrier spacing, and T is symbol spacing in time. The sampling rate is evaluated by $f_s = 1/\Delta t = FN_{FFT}$, where $N_{FFT} \geq L$. Here, N_{FFT} is used to show the FFT size. The sampled basis pulse $g_{l,k}(t)$, can be expressed as vector $g_{l,k} \in C^{N \times 1}$, given by

$$[g_{l,k}]_i = \sqrt{\Delta t}\, g_{l,k}(t)\Big|_{t=(i-1)\Delta t - \frac{OT_o}{2}+T} \quad for\ i = 1,\ldots,N \tag{1.2}$$

with
Next, by combining all the vectors in a matrix $G \in C^{N \times LK}$, given by,

$$N = 0T_0 + T(K-1)f_s \tag{1.3}$$

$$G = [g_{1,1}\cdots g_{L,1} g_{1,2}\cdots g_{L,k}] \tag{1.4}$$

and by representing the combined data vector for all the transmitted symbols as vector $x \in C^{LK \times 1}$,

$$x = vec\left\{\begin{bmatrix} x_{1,1} & \cdots & x_{1,k} \\ \vdots & \ddots & \vdots \\ x_{L,1} & \cdots & x_{L,k} \end{bmatrix}\right\} \tag{1.5}$$

$$x = [x_{1,1}\cdots x_{1,k}\, x_{L,1}\cdots\ldots\ x_{L,k}]^T \tag{1.6}$$

we can write the combined sampled transmit signal $S \in C^{N \times 1}$ as

$$S = G \tag{1.7}$$

The impulse response under multipath fading in a time-variant channel can be expressed as matrix $H \in C^{N \times N}$, whose entries are given by [2],

$$[H]_{i,j} = h[i-j, i] \tag{1.8}$$

Finally, the received symbol is projected using received waveform to obtain

$$y = G^H r = G^H HGx + n \tag{1.9}$$

where n represents the additive noise term which is usually modeled as Gaussian random variable.

1.4   Existing Methods for Equalization in FBMC

1.4.1   One-Tap Zero Forcing Equalizer

In [7], a one-tap zero forcing equalizer for the FBMC system is considered. In this work, it is assumed that self-interference is ignored. Thus, the output of the FBMC system can be formulated as:

$$y \approx diag\{h\} G^H Gx + n \tag{1.10}$$

where h *is* the vector obtain by vectorizing the channel matrix H. Thus, if $\hat{h}$ represents the channel estimate, the output of the one-Tap equalizer will be obtained as:

$$\hat{x} = hardlim(y/\hat{h}) \tag{1.11}$$

Where **hardlim** represents the hard limiter.

1.4.2   MMSE Block Equalizer

The conventional way to design an MMSE equalizer is to use complex-valued symbols. Unfortunately, this does not work with FBMC-OQAM as this equalizer does not eliminate the imaginary interference. A modified

MMSE equalization is formulated to solve this issue by putting real and imaginary parts together using the approach outlined in [11, 13]. This method is termed as full block MMSE equalization whose solution is given by

$$\hat{x} = \begin{bmatrix} Re(\boldsymbol{D}) \\ Im(\boldsymbol{D}) \end{bmatrix}^{T} \left(\begin{bmatrix} Re(\boldsymbol{D}) \\ Im(\boldsymbol{D}) \end{bmatrix} \begin{bmatrix} Re(\boldsymbol{D}) \\ Im(\boldsymbol{D}) \end{bmatrix}^{T} + \Gamma \right)^{-1} \begin{bmatrix} Re(\boldsymbol{y}) \\ Im(\boldsymbol{y}) \end{bmatrix}$$

(1.12)

Where

$$\boldsymbol{D} = \boldsymbol{G}^{H}\boldsymbol{H}\boldsymbol{G} \text{ and } \Gamma = \frac{Pn}{2} \left(\begin{bmatrix} Re(\boldsymbol{G}^{H}\boldsymbol{G}) & -Im(\boldsymbol{G}^{H}\boldsymbol{G}) \\ Im(\boldsymbol{G}^{H}\boldsymbol{G}) & Re(\boldsymbol{G}^{H}\boldsymbol{G}) \end{bmatrix} \right)^{-1}$$

1.5   Proposed Machine Learning-Based FBMC Equalizer

In our proposed equalization scheme, we employ support vector machine (SVM) [16] to learn the required estimate using available input and output data. The SVM is a kind of kernel machine learning technique that utilizes a nonlinear mapping of the original training data [16]. It is mainly designed for binary classification, which was later extended to the multi-class problem. In supervised learning, when labeled training data is inputted, SVM outputs an optimal hyperplane, which categorizes new examples [17, 18]. The variants of SVM used in this study are Linear SVM, Quadratic SVM, and Cubic SVM [19, 20].

The main idea of the proposed equalizer is to learn the weight matrix for the FBMC equalizer via SVM using available training data $\{x_{train}, y_{train}\}$. Once the equalizer weight matrix is learned, the estimate for the unknown transmitted data $x_{unknown}$ is obtained by processing the received signal y_{test} through the designed SVM.

1.6   Results and Discussion

This section provides the results of the proposed machine learning-based equalizers for the FBMC system using LSVM, QSVM, and CSVM.

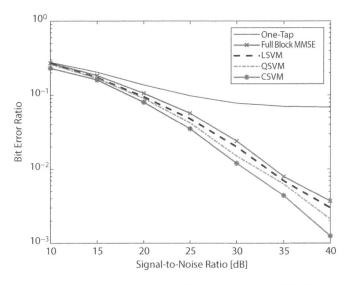

Figure 1.3 BER performance comparison of different equalizers in FBMC system.

Table 1.1 Performance comparison of SVM based FBMC equalizers.

Performance measure	LSVM	QSVM	CSVM
RMSE	0.005951	0.005096	0.0048182
Training Time (s)	86.167	84.591	68.513

In this context, the FBMC system is simulated for 24 subcarriers and 30-time symbols. The prototype filter used is "Hermite" [13]. The results of BER are compared in Figure 1.3, which shows that the proposed SVM based equalizers have better performance than the full block MMSE equalizer of [13]. Moreover, it can be depicted from the results that the CSVM has the best performance among all the proposed methods.

In Table 1.1, the performances of the proposed SVM based equalizers are compared in terms of their testing RMSE and training time in seconds. Again, CSVM is found to be the best among other variants of SVM.

1.7 Summary

In this chapter, we study various methods of FBMC equalizers. We propose machine learning-based equalization techniques for the FBMC system.

For this task, we develop the equalizer using three variants of SVM, namely Linear SVM, Quadratic SVM, and Cubic SVM. The BER performances of these methods are compared with that of the well-known One-Tap and Full Block MMSE equalizers. It was found from the simulation results that the proposed SVM based equalizer has better performance than that of the conventional methods. Moreover, the performances of the SVM based equalizers are compared among themselves in terms of RMSE and training time which shows that the CSVM outperformed the LSVM and QSVM in terms of both the RMSE and training time.

References

1. Farhang-Boroujeny, B., OFDM versus filter Bank multicarrier. *IEEE Signal Process. Mag.*, 28, 3, 92–112, May 2011.
2. Nissel, R., Schwarz, S., Rupp, M., Filter bank multicarrier modulation schemes for future mobile communications. *IEEE J. Sel. Areas Commun.*, 35, 1768–1782, 8, August 2017.
3. Nissel, R., Caban, S., Rupp, M., Experimental evaluation of FBMC-OQAM channel estimation based on multiple auxiliary symbols, in: *IEEE Sensor Array and Multichannel Signal Processing Workshop (SAM)*, Rio de Janeiro, Brazil, July 2016.
4. Nissel, R. and Rupp, M., Enabling low-complexity MIMO in FBMCOQAM, in: *IEEE Globecom Workshops (GC Wkshps)*, Dec 2016.
5. Nissel, R., Z¨ochmann, E., Lerch, M., Caban, S., Rupp, M., Low latency MISO FBMC-OQAM: It works for millimeter waves!, in: *IEEE International Microwave Symposium*, Honolulu, Hawaii, June 2017.
6. Nissel, R., Blumenstein, J., Rupp, M., Block frequency spreading: A method for low-complexity MIMO in FBMC-OQAM, in: *IEEE International Workshop on Signal Processing Advances in Wireless Communications (SPAWC)*, Sapporo, Japan, Jul 2017.
7. Nissel, R. and Rupp, M., OFDM and FBMC-OQAM in doubly-selective channels: Calculating the bit error probability. *IEEE Commun. Lett.*, 26, 6, 1297–1300, 2017.
8. Mestre, X. and Gregoratti, D., Parallelized structures for MIMO FBMCunder strong channel frequency selectivity. *IEEE Trans. Signal Process.*, 64, 5, 1200–1215, 2016.
9. Caus, M. and Pérez-Neira, A.I., Transmitter–receiver designs for highly frequency selective channels in MIMO FBMC systems. *IEEE Trans. Signal Process.*, 60, 12, 6519–6532, 2012.
10. Bellanger, M., FS-FBMC: An alternative scheme for filter bank based multi-carrier transmission, in: *IEEE International Symposium on Communications Control and Signal Processing (ISCCSP)*, 2012.

11. Waldhauser, D.S., Baltar, L.G., Nossek, J.A., MMSE subcarrierequalization for filter bank based multicarrier systems, in: *IEEE SPAWC*, 2008.
12. Ikhlef, A. and Louveaux, J., Per subchannel equalization for MIMOFBMC/OQAM systems, in: *IEEE Pacific Rim Conference on Communications, Computers and Signal Processing*, 2009.
13. Nissel, R., Rupp, M., Marsalek, R., FBMC-OQAM in doubly-selective channels: A new perspective on MMSE equalization. *2017 IEEE 18th International Workshop on Signal Processing Advances in Wireless Communications (SPAWC)*, Sapporo, pp. 1–5, 2017.
14. Marijanović, L., Schwarz, S., Rupp, M., MMSE equalization for FBMC transmission over doubly-selective channels. *2016 International Symposium on Wireless Communication Systems (ISWCS)*, Poznan, pp. 170–174, 2016, doi: 10.1109/ISWCS.2016.7600895.
15. Kumar, A., Albreem, M.A., Gupta, M., Alsharif, M.H., Kim, S., Future 5G Network Based Smart Hospitals: Hybrid Detection Technique for Latency Improvement. *IEEE Access*, 8, 153240–153249, 2020.
16. Kumar, A., Gupta, M., Le, D.N., Aly, A.A., PTS-PAPR Reduction Technique for 5G Advanced Waveforms Using BFO Algorithm. *Intell. Autom. Soft Co.*, 27, 3, 713–722, 2021.
17. Meena, K., Gupta, M., Kumar, A., Analysis of UWB Indoor and Outdoor Channel Propagation. *2020 IEEE International Women in Engineering (WIE) Conference on Electrical and Computer Engineering (WIECON-ECE)*, IEEE, pp. 352–355, 2020.
18. Gupta, M., Chand, L., Pareek, M., Power preservation in OFDM using selected mapping (SLM). *J. Stat. Manage. Syst.*, 22, 4, 763–771, 2019.
19. Lin, G., Lundheim, L., Holte, N., On efficient equalization for OFDM/OQAM systems, in: *Proc. InOWo'05*, Hamburg, Germany, pp. 1–5, Aug. 31–Sep. 1, 2005.
20. Cortes, C. and Vapnik, V., Support-vector networks. *Mach. Learn.*, 20, 3, 273–297, 1995.

Implantable Cardio Technologies: A Review of Integrated Low Noise Amplifiers

P. Vijaya Lakshmi[1], Sarada Musala[1] and Avireni Srinivasulu[2]*

[1]Department of Electronics & Communication Engineering, Vignan's Foundation for Science, Technology and Research (Deemed to be University), Guntur, India
[2]School of Engineering & Technology, K.R.Mangalam University, Gurugram, Haryana, India

Abstract

Wireless medical devices are more in demand due to swift augmentation in the research of remotely diagnosed systems as they provide long term medical assistance. Such devices typically contain a sensor to detect the bio-signals from the patient. But the sensed signals are extremely weak to be further processed and are thus easily susceptible to noise. Hence, a low noise amplifier (LNA) is used post sensor to amplify the weak signal and to reduce the effect of noise. This paper mainly focuses on the LNA suitable for ECG signal acquisition systems to record the electrical motion of a patient's heart. As LNA is the most power thirsty block of the signal acquisition system, this review concentrates on low power LNA suitable for implantable technologies. Since integrated circuit technology provide miniaturization and low power consumption, integrated LNAs are popularly used in the implants. So, the recent integrated low noise low power amplifier designs with high performance and functionality are reviewed mentioning the pros and cons of the designs. Applications of integrated circuit based LNAs are as well referred. The review is concluded with future outlooks and important challenges regarding ECG signal amplification.

Keywords: Low noise amplifier, analog front end, medical implants, integrated circuits, VLSI

*Corresponding author: avireni@ieee.org

Arun Kumar, Manoj Gupta, Mahmoud A. Albreem, Dac-Binh Ha and Mohit Kumar Sharma (eds.)
Wearable and Neuronic Antennas for Medical and Wireless Applications, (11–36) © 2022
Scrivener Publishing LLC

2.1   Introduction

An electrocardiography (ECG) is the most extensively used exploratory tool within cardiology to track the electrical changes on patient's body due to the activities of human heart. This simple measurement indicates heart rhythm and many cardiac problems, like meager blood flow to the heart and structural abnormalities. For this the medical technology designs different equipment that aid in the diagnosis. The key requirement of the ECG signal to be used in diagnostic equipment is amplification because typical ECG signal amplitude ranges between 100 μV and 4 mV [1, 2] and its frequency range lies between 0.1 and 250 Hz [3] as shown in Figure 2.1. So, capturing and analyzing of these signals is difficult because of their small amplitudes. Hence an amplifier is used to amplify the signal and to suppress the noise prior to use in the diagnostic equipment for processing of the signal.

Cardiac recordings can also be obtained using some of the commercially existing amplifier systems [4, 5]. But custom-made ICs offer supreme advantages like miniaturization and low power consumption which are essential for implantable devices. Constant development in novel circuit design techniques and fabrication technology of microelectronics has profited IC bio-potential amplifiers with significant improvement especially with the noise versus power trade-off.

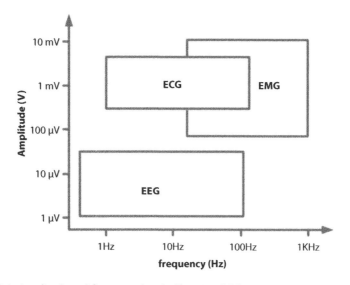

Figure 2.1 Amplitude and frequency band of bio signal [2].

Regardless of the advancements in the low noise amplifiers, designers' thirst towards producing novelty in it has become an exercise for the betterment of the medical technology. For this purpose, researchers should have expertise in circuit design as well as basic knowledge in the biological aspects of the ECG signal. Hence this review comprises background design considerations and concepts of the ECG signal and discusses the state-of-the-art design technologies. The review also discusses the applications of LNA and concludes with future confronts and point of view in the development of the LNA design.

2.2 Background on Low Noise Amplifiers

2.2.1 ECG Signal Characteristics

The characteristics of ECG signals are presented in Figure 2.2 [6]. They are usually small in amplitude around 1 mV and bandwidth less than 1 kHz. So, they require a high gain and low noise amplifier for further processing. Approaching these technical requirements is challenging and requires specialized circuit design skills.

The most challenging point in recording ECG signal is to record high quality signal in the presence of noise. As shown in Figure 2.2 [6] amplitude of the signal to be recorded is very small so, the electronic noise of the

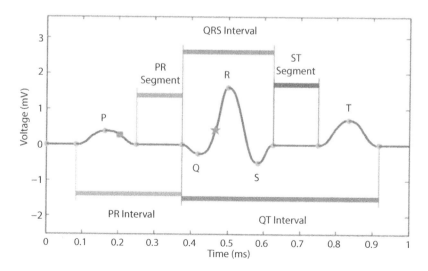

Figure 2.2 ECG signal characteristics [6].

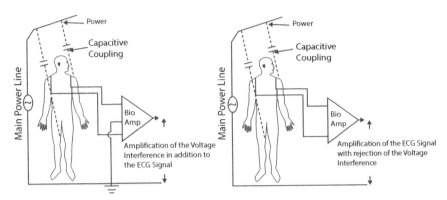

Figure 2.3 Common mode voltages rejection by eliminating the ground electrode [10].

components and the external electrical interference cause the major considerations of noise. Electronic noises of the components include thermal noise and flicker noise. External electrical interference includes baseline wander, power line interference, EMG noise and electrode motion [7, 8]. Baseline wander is the shift in the x-axis of the signal up and down due to the improper electrodes or patient's movement. This can be minimized using narrowband filtering technique. EMG noise originates from the muscle contractions which can be suppressed using moving average filter. Electrode motion noise is caused due to the variation of impedance of the skin around the electrode. This noise is usually cancelled using adaptive filters. Electromagnetic field sourced by power line stands for the common noise source in ECG which is represented by 50 or 60 Hz sine wave with number of harmonics. This noise can be removed using subtraction procedure [9]. These common mode interference noises can also be reduced by eliminating the ground electrode in the setup of ECG signal read out (Figure 2.3) [10]. Consequences of this are high common mode interference which overwhelms the front-end amplifiers, despite a high CMRR [11]. Common mode interference can also be reduced using electrical shielding but practically it is not easy so improving CMRR is desirable. Whereas the electronic noises of the components are reduced with careful designing.

2.2.2 General ECG Readout Amplifier System Architecture and Design Considerations

Figure 2.4 represents the general structure of ECG readout amplifier system. Its analog front-end part consists of LNA which senses the signal

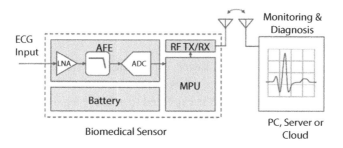

Figure 2.4 Typical configuration of a wearable ECG monitoring system [12].

from electrodes and amplifies it by suppressing noise followed by filter and A/D converter. The signal is then post processed as per the requirement. In the entire process high accuracy and precision are required because small variations in the actual signal would lead to critical diagnostic value.

So, it is expected from the ECG system that any irregularity detected should be due to an unhealthy cardiac action but not because of the equipment used. Hence many special considerations are to be taken into account while designing the LNA. Few important considerations of designing ECG amplifier are [13]:

1. The input impedance of the amplifier should be very high to match the resistance of the sensor.
2. Gain of the amplifier should be high such that the amplified ECG signal could be displayed.
3. It should be able to provide high signal to noise ratio.

Three types of bio-electrodes are used based on the application or requirement for ECG reading referred to as wet, dry, and insulating [14]. Wet are often Ag/AgCl electrodes that make use of an electrolytic gel to form a conduction path between skin and electrode. These electrodes are popularly used in medical applications. Dry electrodes use benign metal (such as stainless steel) with no electrolyte between electrode and skin. Insulating electrodes consist of a metal or semiconductor with a thin dielectric surface layer, so that the bioelectric signal is capacitive coupled from the skin to the substrate. But, for the cardiovascular disease patients, portable ECG is a convenient electronic device. On behalf of the traditional wet gel electrodes used in clinical applications, dry electrodes are preferred for comfort and long-time wearing. But the source impedance of these dry electrodes is even higher than the wet electrodes. Hence a high input impedance amplifier is preferred to match with the impedance of the electrodes [15]. Results with 99% of data analysed show that ECG recording amplifier with an input impedance of 500 MΩ must guarantee a distortion less sensed signal [16].

This amplifier must have lower input-referred noise and high gain because noise of the next stages refers to the electrodes through this amplifier's gain. Usually gain of LNA is expected to achieve a gain of about 40–60 dB to be able to display the signal [17]. The succeeding one or more amplifier stages provide additional gain and limit the bandwidth. The choice of gain is always limited by the application. Finally, obtained amplified signal is then processed.

One of the important requirements of ECG acquisition systems is the potential to detect very small input signals at near DC frequencies, where flicker is overrides. For this amplifier should be designed such that its input referred noise is less than 3 μV_{rms} as per the survey from the state-of-the-art amplifiers, so that the inherent noise of the amplifier is less than the noise of the electrodes.

In the recent years, there is a strong demand for active implants in the medical industry. So, apart from the three considerations mentioned for designing ECG amplifier, an implantable technology demands a low power ECG amplifier for the purpose of extended battery life. At a cost of all the technical constraints the implants are powered through battery that is expected to be reliable, capable and to have more active life [18]. As well the technical designers are also actively participating to implement low power design techniques to provide low power circuits for prolonged usage of the battery span. This paper concentrates on low power and low noise amplifier designs for ECG signal read outs.

2.2.3   Low Noise Amplifier Circuit Design

Designing low noise amplifier circuit for Figure 2.4 needs exceptional care because performance of this amplifier determines the accuracy of the entire front-end system. But most of these amplifiers suffer with high values of input referred noise, power consumption and limited bandwidth. So, a range of closed loop amplifier topologies like capacitive coupled capacitive feedback, direct coupled active feedback and chopper stabilised are proposed to prevail over the mentioned problems but, each having their own merits and demerits. The amplifier topology most popularly used is, capacitive coupled capacitive feedback topology as mentioned in Figure 2.5 for acquisition of most of the biomedical readings [19–25]. Fully differential structure is utilised to ensure high common mode rejection ratio (CMMR) and power supply rejection ratio (PSRR).

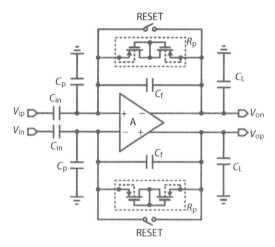

Figure 2.5 Fully differential capacitive coupled feedback amplifier [19].

In this structure capacitors are used to set the mid-band gain (C_{in}/C_f) and to reject DC offset from the skin to electrode. Gain of the amplifier is expected to be high to suppress the noise. This can be achieved by increasing the value of C_{in} but at a cost of silicon area. So, the amplifier block used in the topology is circumspectly designed to obtain higher gain. The high pass corner frequency is set by the feedback resistor and capacitor. Feedback resistor is often a pseudo resistor formed with pMOS transistors that aid in establishing proper DC feedback bias. Low pass corner frequency is defined by the amplifier itself using feedback capacitor, based on the usable spectral data in the typical ECG signal.

Noise efficiency factor (NEF) is one of the prime performances metric to compare different front-end amplifier topologies. NEF signify number of times the noise of a system with same bias current and bandwidth is higher compared to the ideal case [26]. It is represented as

$$\text{NEF} = V_{\text{rms, in}} \sqrt{\frac{2.I_{tot}}{\pi.U_T.4KT.BW}} \qquad (2.1)$$

where $V_{\text{rms,in}}$ is total equivalent input noise, I_{tot} is the total bias current of the amplifier, U_T is the thermal voltage, K is the Boltzmann's constant, T is the absolute temperature and BW is the amplifier's effective noise bandwidth. But as this prominent metric does not include supply voltage, comparison of amplifiers operated at different supply voltages is not possible. Hence power efficiency factor (PEF) [27] is introduced which includes both operating current and supply voltage therefore providing a better comparison between the bio signal acquisition amplifiers. It is represented as

$$PEF = NEF^2.V_{dd} \qquad (2.2)$$

The input referred noise is one of the most important specifications of ECG read out amplifier and as per [28] the relation between overall input referred noise of the capacitive feedback amplifier and the input referred noise of the amplifier used is as mentioned in Eq. (2.1).

$$\overline{v_{nl,overall}^2} = \left(\frac{Cin + Cp + Cf}{Cin} \right)^2 \overline{v_{nl,A}^2} \qquad (2.3)$$

To provide low noise performance, LNA block of analog front end (AFE) consumes the maximum power. Hence design care is taken with the amplifier used in the LNA to obtain a trade-off between power and noise, as both are significant parameters for the implants. So, a capable amplifier with high gain, low noise and low power is critical to design to be used in this topology.

2.2.4   Operational Transconductance Amplifier Circuits Used in LNA Design

Operational transconductance amplifier is an operational amplifier without a buffer. It is a voltage controlled current source that is it provides current proportional to an input voltage difference. Since current is amplified rather than voltage, design is simple and has minimized 1/f noise effect. So, OTA is chosen by many designers for the amplification of bio signals in the analog front end. OTA's are as well suitable for generating signals too [29–31].

There are different forms of OTA's available like, Single-stage OTA, Two-stage OTA, Folded cascode OTA, and Telescopic OTA, each having their pros and cons. Single stage OTA has the advantage of simplicity in design and high speed but because of its low output impedance it has relatively low gain. Two-stage OTA as the name depicts has two stages, in which one provides high gain followed by the stage that offers high voltage swing. It overcomes the disadvantage of Single stage OTA but at a cost of area, power consumption, reduced speed, and compromised frequency response. Folded cascode amplifier has superior frequency response compared to two-stage OTA but, as it uses a greater number of devices comparatively it contributes more significant input referred thermal noise to the signal and achieves lesser gain compared to two-stage. The telescopic architecture generates a more direct signal path by arranging input differential pair and output branch of folded cascode on the same current branches. This brings the advantage of simplicity, high speed, and reduction in the usage of the bias current [32, 33]. However, it has a limited output swing.

Furthermore, different new techniques are implemented to these existing OTA's to increase gain along with trade-offs. Few important techniques are:

1. Gain boosting technique: It offers higher DC gain than two-stage and low power consumption compared to three-stage OTA. It also provides a rail-to-rail output voltage swing but with reduced speed [34].
2. Current boosting technique: Second stage of two-stage amplifier is designed using class-AB, to achieve large values of gain, bandwidth and slew rate [35, 36].
3. Enhancing input common mode range: This is done using complementary input configuration applied to a common mode feedback circuit (CMFB), which suppresses the common mode signal components and stabilises the common mode voltage at high impedance nodes [37].
4. Miller compensation: Most of the designs incorporate two-stage OTA, usually a two-pole system. To provide good phase margin and ensure stability, these poles are separated by moving dominant pole to low frequency and non-dominant pole to higher frequency. This is accomplished using miller capacitance which can easily convert low capacitance into equivalent high capacitance value with Miller effect.
5. Inverter based differential input stage: This is widely used conceptual structure to provide two trans-conductors with only one branch of DC current [20].
6. Current reuse technique: It is technique used for reducing the power dissipation in the low noise amplifier by not using an extra current source for biasing the load instead using current generated by the driver transistor [38]. But this doubles the gate capacitance of the input differential pair.

So, it can be inferred from these discussions that, inspite of limited output voltage swing, telescopic cascode OTAs are popularly used in reading bio-signals because of their high gain, area efficiency, low power, and low noise benefits. The state-of-the-art telescopic OTAs used for reading ECG signals, use two-stage to overcome the limited output swing and complimentary differential pairs along with current reuse technique to double the trans-conductance using the same current consumption. They are:

2.2.4.1   Typical Telescopic Cascode Amplifier [39]

The design Figure 2.6 makes use of telescopic cascode rather than folded cascode to reduce the number of active twigs. As thermal noise is dominant than flicker noise in this design, differential input is applied to nMOS pair rather than pMOS to achieve higher g_m/I_D ratio. Thermal noise and flicker noise of this design are represented in Eqs. (2.4) and (2.5) respectively.

$$v_{ni,th}^2 = \left(\frac{16kT}{3g_{m1}} \left(1 + \frac{g_{m5}}{g_{m1}} \right) \right) \Delta f \tag{2.4}$$

$$v_{ni,\frac{1}{f}}^2 = \frac{1}{C_{ox}\Delta f} \left(\frac{K_n}{(WL)_1} + \frac{K_p g_{m5}^2}{(WL)_5 g_{m1}^2} \right) \tag{2.5}$$

So, when the circuit is carefully designed with large size input transistors and reduced g_{m5}/g_{m1} ratio, achieves low noise. Even though it attains good noise-power performance, additional current is consumed with the differential input branches and the second stage.

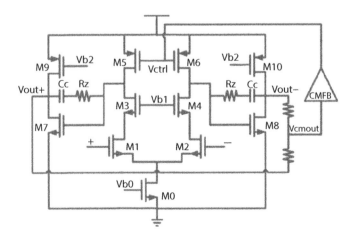

Figure 2.6 Typical telescopic cascode amplifier [39].

2.2.4.2 Complementary Input Closed Loop Amplifier [40]

This design is analogous to the preceding design but, to reduce the input referred noise significantly, both nMOS and pMOS differential pairs were driven by the input signal as shown in Figure 2.7. The complementary input method also doubles the transconductance of the amplifier as it is inverter based differential input stage technique. The input referred thermal noise and flicker noise of this design considering equal gm1 and gm3 are as in Eqs. (2.6) and (2.7) respectively.

$$v_{ni,th}^{2} = \left(\frac{16kT}{3g_{m1}.2} \right)\Delta f \qquad (2.6)$$

Consequently, the proposed structure offers higher DC gain comparing to the two-stage and gain-boosting structures with much lower power consumption comparing to the three-stage OTA. Also, the same as other mentioned OTA structures, it provides a rail-to-rail output voltage swing. Consequently, the proposed structure offers higher DC gain comparing to the two-stage and gain-boosting structures with much lower power consumption comparing to the three-stage OTA. Also, the same as other mentioned OTA structures, it provides a rail-to-rail output voltage swing. Consequently, the proposed structure offers higher DC gain comparing to the two-stage and gain-boosting structures with much lower power

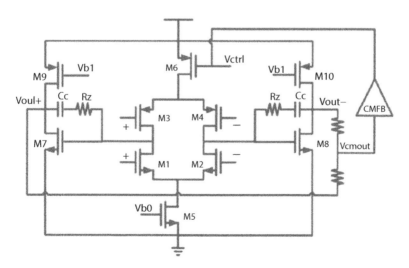

Figure 2.7 Complementary input closed loop amplifier.

consumption comparing to the three-stage OTA. Also, the same as other mentioned OTA structures, it provides a rail-to-rail output voltage swing. Consequently, the proposed structure offers higher DC gain comparing to the two-stage and gain-boosting structures with much lower power consumption comparing to the three-stage OTA. Also, the same as other mentioned OTA structures, it provides a rail-to-rail output voltage swing.

$$v^2_{ni,\frac{1}{f}} = \frac{1}{4C_{ox}\Delta f}\left(\frac{K_n}{(WL)_1} + \frac{K_p}{(WL)_3}\right) \tag{2.7}$$

Thermal noise and flicker noise of this design will be $1/\sqrt{2}$ times the previous design provided, in Figure 2.6 if $g_{m5} = 0$ and $k_n/(WL)_1 = K_p/(WL)_3$. Hence this design achieves good noise-power trade off compared to the previous one. As trans-conductance doubles, to overcome the increase in the common mode gain, dual tail current sources are added in the first stage, to degenerate common mode trans-conductance and increase CMRR/PSRR. This efficient design still draws some considerable amount of current in the first stage from supply, as mentioned in the comparative analysis of [40].

2.2.4.3   Fully Differential Current Reuse OTA [41]

To get better with the current efficiency this design in Figure 2.8 employs current reuse technique at the input stage. The four input transistors are

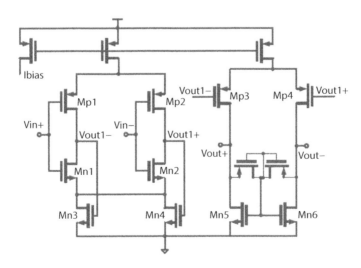

Figure 2.8 Fully differential current reuse OTA [41].

made to work in the weak inversion region to reduce the thermal noise and large gate areas are preferred to minimise the flicker noise. M_{n3}, M_{n4} and M_{n5}, M_{n6} provide CMFB to the input stage and output stage, respectively. Output CMFB also includes large resistance realised by two diode connected transistors to limit the current consumption of the output stage. The thermal noise of this design is depicted as in Eq. (2.8).

$$v_{ni,th}^2 = 2 * \left(\frac{4kTn}{(g_{mn1} + g_{mp1})} \left(\Gamma_{wi} + \frac{\Gamma_{si} g_{mn3} + \Gamma_{wi} g_{mp3}}{g_{mn1} + g_{mp1}} \right) \right) \Delta f \quad (2.8)$$

Where Δf is the bandwidth, g_{mni} and g_{mpi} indicates the transconductance of nMOS transistor M_{ni} and pMOS transistor M_{pi}. Γ_{wi} and Γ_{si} represent the thermal noise for weak inversion and strong inversion, respectively.

This simple design draws very less bias current from the supply which is the advantage compared to the previous designs. This gives a low NEF value. Even though the circuit provides double transconductance due to the implemented current reuse technique but still its input referred noise is greater than 5 μV_{rms}, which must be reduced.

2.2.4.4   Fully Reconfigurable OTA Using Floating Gate Transistors [42]

This design presents a reconfigurable bio-signal readout amplifier like previous mentioned designs additionally using floating gate transistors. Floating gate transistors are employed for common-mode feedback, bias generation, and programmable feedback resistances to avail the advantage of ultra-low power and ultra-low voltage.

Transconductance of the differential pair M_{1-4} in Figure 2.9 is set by the bias current I_b, which is sourced from the floating gate transistor that is programmed. Common mode feedback is implemented using a pMOS floating gate transistor M_{cmfb}. The charge Q_{cmfb} on this transistor can be programmed to adjust the floating gate voltage V_{fg}. The output common mode voltage remains constant when the Q_{cmfb} is invariant. Pseudo resistors (not shown in the circuit) are also realized using floating gate-based transistors to reconfigure the low frequency corner cut off so that the design can sense various electrophysiological signals.

Using complimentary differential pairs and floating gate CMFB the design achieves good trade-off between noise and power. The input referred noise and NEF of the design are low. The practical problem with floating gate-based OTA is its silicon area because C_1 and C_2 capacitors are maintained to

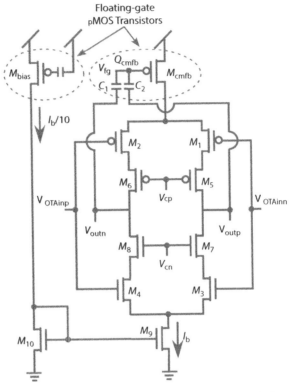

Figure 2.9 Fully reconfigurable OTA using floating gate transistors [42].

be 10 times larger than the oxide capacitance of the floating gate for proper functioning of the device [43]. Also, the NEF can be furthermore reduced by reducing the current consumption in the biasing branch by implementing common mode feedback with nMOS floating gate devices.

2.2.4.5   Low Noise OTA with Output Boosting Technique [44]

As with previous designs this circuit in Figure 2.10, also avails the advantages of the two-stage, inverter based fully differential pair and current reuse techniques. The input transistors are biased in the subthreshold region, to improve the noise to power efficiency. The input referred thermal noise of this design is approximately

$$v_{ni,th}^2 = \left(\frac{4kTn}{g_{mi}} \left(1 + \frac{g_{m7,8}}{g_{mi}} + \frac{g_{m9,10}}{g_{mi}} \right) \right) \Delta f \tag{2.9}$$

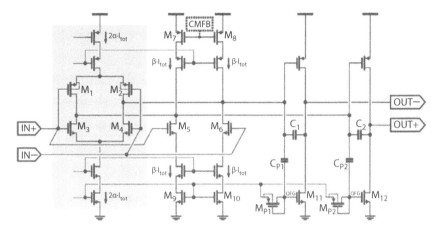

Figure 2.10 Low noise OTA using quasi floating gate (QFG) transistors in the output stage [44].

CMFB in this design requires minimum current dissipation for common mode settling and loop stability. To lessen the output distortion, large slew rate is required. For this linearity has to be improved by increasing static current in the output stage. But more current in the output stage will increase the power factor. To compensate an output boosting technique using quasi floating gate transistors are employed to push the operation of second stage into class-AB. This technique also increases the transconductance compared to the floating gate transistors. But the quasi floating gate transistors used reduce the gain bandwidth and output impedance of the second stage [45]. Also, it utilizes large area as it uses large capacitance to pass a small signal.

2.2.4.6   Low Noise Low Power OTA [46]

Figure 2.11 is a low noise and low power OTA with current reused technique not including bias circuit and CMFB amplifier. It employs inverter-based input stage for low noise and class-AB output stage for large output range and high g_m/I efficiency. In this design current reuse technique is implemented by merging the driving branch of the class-AB stage into the input stage to reduce the power consumption. This facilitates the use of supply current both in the input and output stage to achieve high linearity and noise-power efficiency. This design concentrates on the driving circuit of the class-AB stage to stabilise the quiescent current in the output stage, while providing equal small signal components to both the transistors of the output stage. The other techniques existing for realising the driving

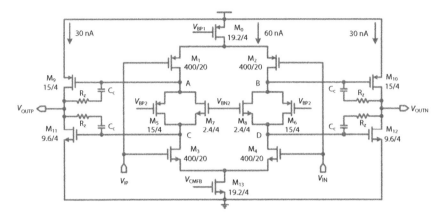

Figure 2.11 Low noise low power OTA [46].

circuit like current mirror biasing, floating controlled class-AB, etc. use additional current consuming branch. Realizing the driving circuit without extra current consumption can also be implemented using QFG technique as mentioned in the previous design but it uses more silicon area as it needs a large capacitance for passing a small signal. So, this circuit uses a control unit made up of $M_5(M_6)$ and $M_7(M_8)$ transistors that are inserted in the first stage in between the inverter based differential pair so as to use the DC current of the first stage. The input referred thermal noise of this design is represented as

$$v_{ni,th}^2 = \left(\frac{4kTn}{g_{m1,2} + g_{m3,4}} \right) \tag{2.10}$$

Even though the design achieves good NEF among the available two-stage OTA but, its PEF is not the best one. The minimum supply voltage of the design is limited to obtain required dynamic range. Also, the CMFB block implemented utilizes large silicon area.

2.2.4.7 Cross Coupled Load Current Reuse OTA [47]

The main concept of this design is to improve CMRR. As cross coupled loads usually reduce the gain, both differential and common mode gain reduces implementing this concept. But to satisfy the requirement of high differential gain, current reuse fully differential amplifier is used. So, this design Figure 2.12 adds cross coupled load to such amplifier to achieve high differential gain and attenuated common mode gain.

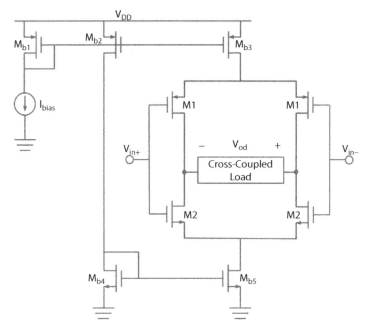

Figure 2.12 Cross coupled load current reuse OTA [47].

If there is no loading effect of the cross coupled load, then CMRR of this design can be increased up to 30 dB than the usual differential current reuse amplifier with common mode feedback. But this technique of improving CMRR would be more impactful if it does not reduce the differential gain of the amplifier.

2.2.4.8   Fully Differential Stacked OTA [48]

Unlike the rest of the designs mentioned, this circuit is realised in chopper stabilized amplifier rather than capacitive feedback. This is still considered for review in the group of capacitive coupled feedback because of its stacking concept. Figure 2.13 stacks up multiple OTA stages on one another where the transconductance of the overall design will be the sum of the individual OTA transconductance. Individual stage is implemented using inverter-based input stage to avail the advantage of doubled transconductance and low noise. Stacking reduces input referred noise and achieves good NEF and PEF values. Thermal noise and flicker noise of this stacked design expressed in Eqs. (2.11) and (2.12) respectively convey that noise is inversely proportional to the number of stacked stages N.

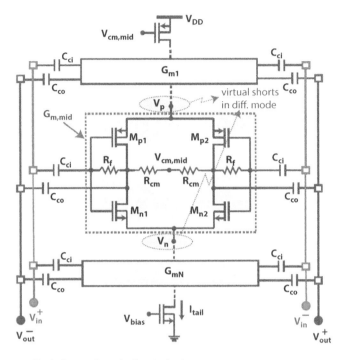

Figure 2.13 Fully differential stacked OTA [48].

$$v_{ni,th}^2 = \left(\frac{4k_B T \gamma}{2N g_m} \right) \tag{2.11}$$

$$v_{ni,\frac{1}{f}}^2 = \frac{1}{C_{ox} 4Nf} \left(\frac{K_n}{(WL)_n} + \frac{K_p}{(WL)_p} \right) \tag{2.12}$$

Another advantage of this design is it implements self-biased inverters at the input stage. Self-bias method facilitates to maintain a good CMRR and is accomplished with a feedback pseudo resistor R_f. Drawbacks of the design OTA stacking are increase in area, supply voltage, and input parasitic capacitance.

Therefore, all the mentioned amplifiers so far use the advanced techniques to obtain low NEF and PEF values with low power consumption and to be operable at low supply voltages. Table 2.1 mentions the comparative analysis of the amplifiers discussed. Among all these designs [41, 44, 46, 48] structures employ current reuse techniques effectively and utilize the bias current efficiently to the most possible branches. Hence these designs achieve

Table 2.1 Performance summary of the discussed low noise amplifiers.

	[39]	[40]	[41]	[41]	[44]	[46]	[47]	[48]
V_{dd} (V)	1	1	1	2.5	2.4	2	1.8	0.95
Gain (dB)	40.5	38.3	34	40.74	47	39.8	46	36
I_{amp} (μA)	12.5	12.5	0.8	4.3	1.46	0.16	10	0.013
$V_{ni,rms}$ (μV)	3.2	1.95	5.71	2.8	3.31	2.05	0.43	0.194
NEF	4.5	2.48	2.59	2.25	2.4	2.26	–	1.08
PEF	20.3	6.15	6.71	–	–	10.2	–	1.12
CMRR (dB)	60	63	60	70	65	65	98	95
PSRR (dB)	60	63	70	70	76	70	–	68
Bandwidth	0.4–8.5k	0.023–11.5k	0.2–430	0.1–10k	250	0.2–200	4.8k	240
Area (mm²)	0.047	0.072	–	0.17	–	0.18	–	0.18
Technology (μm)	0.13	0.13	0.18	0.35	0.35	0.35	0.18	0.18

low power consumption comparatively because of their reduced amount of utilization of bias current [47, 48] have lesser input referred noise and high CMRR because of the special techniques implemented, among which [47] would be better choice with less area. All the designs are chosen such that their NEF value is less such that they are suitable for the ECG readouts. Even though all these designs satisfy the requirements of the ECG readout, the best can be chosen based on the requirement.

2.3   Applications of Low Noise Amplifiers

Low noise amplifiers (LNA) have wide variety of applications in biomedical sector because of their ability to read small signals in the noisy environment. They are used in (i) biosensors implanted in human body to detect and analyse the bio signals, (ii) for measuring and recording ECG signals and (iii) neural recording systems that are interfaced with wireless transmission for measuring neural activity. This paper concentrates mainly on the amplifiers able to deal with the ECG signals, so the focus is on applications (i) and (ii).

2.3.1   For Implantable Bio-Sensors

With the increase in demand for the biological information requirement in medical diagnostics and health care, implantation of bio sensors has been increasing. Most of the implantable devices related to ECG signal are used to monitor or even to control the irregular heartbeats in patients suffering with heart rhythm disorders. These devices include loop recorders, pacemakers, and implantable cardioverter defibrillators.

Loop recorder is an implanted wireless cardiac monitor used for recording heart's rhythm for up to three years. Patient's with unexplained heart palpitations or fainting spells that cannot be detected in short term heart's electrical activity recording are suitable for implanting this device. Pacemaker is a device which is implanted under the skin of the chest to produce electrical pulses to maintain a normal rate heartbeat. That is it helps to manage heart rhythm disorders such as arrhythmia when heart beats irregularly or bradycardia when heart beats too slowly. An ICD detects irregular heartbeat and automatically shock's the heart rhythm back to normal.

In all the three implantable devices mentioned the primary job is to observe the heart rhythm ECG. Hence a low noise amplifier is the very first block of each of these devices for amplifying the sensed signal and providing it for the further processing [49]. This paper mainly projects amplifiers suitable for this application.

2.3.2   For Measuring and Recording ECG Signal

ECG signal is a graph that records variations in the electric potential changes with the patient's torso to demonstrate the cardiac activity. It mainly tracks the heart rhythm through which it further discloses the cardiac diseases. As the contractions of the heart wall spread electrical currents from the heart throughout the body, action potentials are created. These potentials are sensed by electrodes placed on the skin. From these electrodes' ECG signal is acquired using a low noise amplifier. This is rather a clinical application for monitoring, recording, analysing and even for providing therapeutic solutions.

2.4   Conclusion

The ECG amplifier is an essential block of analog front end for sensing ECG signal and fleeting it for further processing. This review discusses the essentials of the ECG signal and its characteristics for designing an appropriate amplifier to read it. As it is the first block of wearable ECG monitoring system, efficiency of the entire system depends on it. Hence a high gain, low noise and low power amplifier is looked-for. Closed loop topologies of implementing such an amplifier are mentioned among which capacitive coupled feedback is considered for its simplicity. The types of low noise amplifiers usually used in the topology are then discussed. Among them telescopic cascoded amplifier is better for the application chosen. Inverter based differential pair is implemented to double the transconductance and current reuse technique is used for reducing the consumption of bias current. So, state-of-the-art telescopic inverter based differential inputs with current reuse technique circuits are reviewed and discussed. Each design uses additional technique for reducing its input referred noise and power consumption thus making itself even more suitable for the application of implantable ECG devices.

References

1. Webster, J.G., Medical Instrumentation: Application and Design, Wiley, New York, 1995.
2. Song, H., Park, Y., Kim, H., Hyoungho, K., Fully Integrated Biopotential Acquisition Analog Front-End IC. *J. Sensors*, 15, 25139–25156, 2015. doi:10.3390/s151025139.

3. Tsai, T.-H., Hong, J.-H., Wang, L.-H., Lee, S.-Y., Low-Power Analog Integrated Circuits for Wireless ECG Acquisition Systems. *IEEE Trans. Inf. Technol. Biomed.*, 16, 5, 907–917, 2012. doi:10.1109/titb.2012.2188412.
4. Boston Scientific. [Online]. Available: http://www.bostonscientific.com/products.
5. Medtronic. [Online]. Available: http://www.medtronic.com/products.
6. Kovács, P., *Transformation Methods in Signal Processing*, 2016, doi:10.15476/ELTE.2015.187.
7. Matteo, D.A., Longo, A., Rizzi, M., Noisy ECG signal analysis for automatic peak detection. *eHealth Artif. Intell.*, 10, 2, 35, 2019. https://doi.org/10.3390/info10020035.
8. Rahul, K., Signal Processing Techniques for Removing Noise from ECG Signals. *J. Biomed. Eng. 1*, 3, 101, 1–9, 2019.
9. Chavdar, L., Georgy, M., Ratcho, I., Ivan, D., Ivaylo, C., Dotsinsky, I., Removal of Power Line Interface from the ECG: A Review of the Subtraction Procedure. *Biomed. Eng. Online*, 4, 50, 1–18, 2005. doi: 10.1186/1475-925X-4-50.
10. Hamza, N., Khriji, L., Tourki, R., Interference Reduction in ECG Signal Acquisition: Ground Electrode Removal (case study). *International Conference on Computer Medical Applications (ICCMA)*, 2013. doi:10.1109/iccma.2013.6506165.
11. Metting, V.R.A.C., Peper, A., Grimbergen, C.A., High-quality recording of bioelectric events. Part 1. Interference reduction, theory and practice. *Med. Biol. Eng. Comput.*, 28, 5, 389–397, 2005.
12. Hirai, Y., Matsuoka, T., Tani, S., Isami, S., Tatsumi, K., Ueda, M., Kamata, T., A Biomedical Sensor System with Stochastic A/D Conversion and Error Correction by Machine Learning. *IEEE Access*, 7, 21990–22001, 2019. doi: 10.1109/ACCESS.2019.2898154.
13. Rowlands, D., James, D., Vanegas, C., Rao, S., Lisner, P., Design and fabrication of an ECG amplifier on silicon using standard CMOS process. *Proc. of IEEE Sensors*, 2003. doi:10.1109/icsens.2003.1279167.
14. Searle, A. and Kirkup, L., A direct comparison of wet, dry and insulating bioelectric recording electrodes. *Physiol. Meas.*, 21, 2, 271–283, 2000. doi:10.1088/0967-3334/21/2/307.
15. Weilin, X., Taotao, W., Xueming, W., Hongwei, Y., Baolin, W., Jihai, D., Haiou, L., Low Noise, High Input Impedance Digital-Analog Hybrid Offset Suppression Amplifier for Wearable Dry Electrode ECG Monitoring. *Electronics*, 9, 165, 1–13, 2020. doi: 10.3390/electronics9010165.
16. Assambo, C. and Burke, M.J., Amplifier input impedance in dry electrode ECG recording. *Annual International Conference of the IEEE Engineering in Medicine and Biology Society*, 2009. doi:10.1109/iembs.2009.5333398.
17. Kwak, J., *A Low-noise Amplifier for Electrocardiogram Signals. Master's thesis*, University of California, Berkeley, 2015.
18. Ben Amar, A. *et al.*, Power Approaches for Implantable Medical Devices. *Sensors Basel*, 15, 11, 28889–914, 2015. doi:10.3390/s151128889.

19. Harrison, R.R., The design of integrated circuits to observe brain activity. *Proc. IEEE*, vol, 1203–1216, 2008.

20. Deepu, C.J., Zhang, X., Liew, W.S., Wong, D.L.T., Lian, Y., AnECG-on-chip with 535 nW/channel integrated lossless data compressor for wireless sensors. *IEEE J. Solid-State Circuits*, 49, 11, 2435–2448, Nov. 2014.

21. Wang, T.Y., Lai, M.R., Twigg, C.M., Peng, S.Y., A fully reconfigurable low-noise biopotential sensing amplifier with 1.96 noise efficiency factor. *IEEE Trans. Biomed. Circuits Syst.*, 8, 3, 411–422, Jun. 2014.

22. Zhang, F., Holleman, J., Otis, B.P., Design of ultra-lowpower biopotential amplifiers for biosignal acquisition applications. *IEEE Trans. Biomed. Circuits Syst.*, 6, 4, 344–355, Aug. 2012.

23. Chen, Y.P. *et al.*, An injectable 64 nW ECG mixed-signal SoC in 65 nmfor arrhythmia monitoring. *IEEE J.Solid-State Circuits*, 50, 1, 375–390, Jan. 2015.

24. Song, S. *et al.*, A low-voltage chopper-stabilized amplifier for fetal ECG monitoring with a 1.41 power efficiency factor. *IEEE Trans. Biomed. Circuits Syst.*, 9, 2, 237–247, Apr. 2015.

25. Yazicioglu, R.F., Kim, S., Torfs, T., Kim, H., Hoof, C.V., A 30 µW analog signal processor ASIC for portable biopotential signalmonitoring. *IEEE J. Solid-State Circuits*, 46, 1, 209–223, Jan. 2011.

26. Steyaert, M.S.J. and Sansen, W.M.C., A micropower low-noise monolithic instrumentation amplifier for medical purposes. *IEEE J. Solid-State Circuits*, 22, 6, 1163–1168, 1987. doi:10.1109/jssc.1987.1052869.

27. Muller, R., Gambini, S., Rabaey, J.M., A 0.013 mm2, 5 µW, DC-Coupled Neural Signal Acquisition IC With 0.5 V Supply. *IEEE J. Solid-State Circuits*, 47, 1, 232–243, 2012. doi:10.1109/jssc.2011.2163552.

28. Harrison, R.R. and Charles, C., A low-power low-noise CMOS amplifier for neural recording applications. *IEEE J. Solid-State Circuits*, 38, 6, 958–965, Jun. 2003.

29. Bhargav, A., Srinivasulu, A., Pal, D., An Operational Transconductance Amplifiers Based Sinusoidal Oscillator Using CNTFETs, in: *Proc.of the 23rd IEEE International Conference on Applied Electronics (IEEE ICAE-2018)*, Pilsen, Czech Republic, p. 6, 11 Sept–13 Sept, 2018, doi: 10.23919/AE.2018.8501428.

30. Srinivasulu, A., Sowjanya, G., Gautham, S.H., Pitchaiah, T., Krishna, V.V.S.V., Chapter No: 28 Operational Transconductance Amplifiers based sinusoidal oscillator with grounded capacitors, in: *Lecture Notes in Electrical Engineering (LNEE)*, vol. 403, pp. 343–352, Springer, Singapore, 2017, https://link.springer.com/chapter/10.1007/978-981-10-2999-8_28. doi: 10.1007/978-981-10-2999-8_28.

31. Srinivasulu, A., Tejaswini, V., Pitchaiah, T., Time marker generator using Operational Trans-conductance Amplifier. *Jurnal Teknologi*, 76, 1, 103–105, 2015, E-ISSN:2180-3722. doi: 10.11113/jt.v76.4060.

32. López-Martín, A.J., Low-Voltage Super Class AB CMOS OTA Cells With Very High Slew Rate And Power Efficiency. *IEEE J. Solid-State Circuits*, 40, 5, 1068–1077, May 2005.

33. Tianwang, L., Bo, Y., Jinguang, J., A Novel Fully Differential Telescopic Operational Transconductance Amplifier. *J. Semicond.*, 30, 8, August 2009.

34. Najjarzadegan, M., Jalili, A., Dehghani, R., *A High-Speed, High-Gain OTA Structure With A New Compensation Technique*, 2015. doi: 10.1109/IranianCEE.2015.7146423.

35. Noormohammadi, M., Lazarjan, V.K., HajSadeghi, K., New Operational Transconductance Amplifiers using current boosting. *2012 IEEE 55th International Midwest Symposium on Circuits and Systems (MWSCAS)*, doi:10.1109/mwscas.2012.6291969.

36. Sivakumari, K., Srinivasulu, A., Venkata Reddy, V., A high slew rate, low voltage CMOS Class-AB amplifier, in: *Proc. of IEEE Applied Electronics 2014 International Conference (IEEE AEIC-14)*, Pilsen, Czech Republic, pp. 267–270, Sep 9–10, 2014. doi: 10.1109/AE.2014.7011717.

37. Hati, M.K. and Bhattacharyya, T.K., Design of a Low Power, High Speed Complementary Input Folded Regulated Cascode OTA for a Parallel Pipeline ADC. *2011 IEEE Computer Society Annual Symposium on VLSI*, doi:10.1109/isvlsi.2011.9.

38. Chandrashekar, K. and Bakkaloglu, B., A 10 b 50 MS/s Opamp-Sharing Pipeline A/D With Current-Reuse OTAs. *IEEE Transactions on Very Large Scale Integration (VLSI) Systems*, vol. 19, pp. 1610–1616, 2011. doi: 10.1109/tvlsi.2010.2052376.

39. Zhang, F., Holleman, J., Otis, B.P., Design of Ultra-Low Power Biopotential Amplifiers for Biosignal Acquisition Applications. *IEEE Trans. Biomed. Circuits Syst.*, 6, 4, 344–355, 2012. doi: 10.1109/tbcas.2011.2177089.

40. Rai, S., Holleman, J., Pandey, J., Zhang, F., Otis, B., Design of Ultra-Low Power Biopotential Amplifiers for Biosignal Acquisition Applications. 213a, 212–213. Aug.–Dec. 2009.

41. Liu, Zou, X., Goh, W.L., Ramamoorthy, R., Dawe, G., Je, M., 800 nW 43 $nV/\sqrt{Hz}$ neural recording amplifier with enhanced noise efficiency factor. *Electron. Lett.*, 48, 9, 479–480, 2012. doi: 10.1049/el.2012.0685.

42. Wang, T.-Y., Lai, M.-R., Twigg, C.M., Peng, S.-Y., A Fully Reconfigurable Low-Noise Biopotential Sensing Amplifier With 1.96 Noise Efficiency Factor. *IEEE Trans. Biomed. Circuits Syst.*, 8, 3, 411–422, 2014. doi:10.1109/tbcas.2013.2278659.

43. Urban, C.S., *Scaling the Bulk-Driven MOSFET into Deca-Nanometer Bulk CMOS Technologies*, Wallace Library of the Rochester Institute of Technology, Rochester, 2011.

44. Deepu, C.J., Zhang, X., Liew, W.-S., Wong, D.L.T., Lian, Y., An ECG-on-Chip With 535 nW/Channel Integrated Lossless Data Compressor for Wireless Sensors. *IEEE J. Solid-State Circuits*, 49, 11, 2435–2448, 2014. doi:10.1109/jssc.2014.2349994.

45. Khateb, F., Dabbous, S.B.A., Vlassis, S., A survey of non-conventional techniques for low-voltage low-power analog circuit design. *Radioengineering*, 22, 2, 415–427, 2013.

46. Zhang, J., Zhang, H., Sun, Q., Zhang, R., Low-Noise, A., Low-Power Amplifier With Current-Reused OTA for ECG Recordings. *IEEE Trans. Biomed. Circuits Syst.*, 12, 3, 700–708, 2018. doi:10.1109/tbcas.2018.2819207.

47. Habibzadeh, E., Molinas, M., Ytterdal, T., Modified Current-resuse OTA to Achieve High CMRR by utilizing Cross-coupled Load. *PRIME 2019*, Switzerland, doi: 10.1109/PRIME.2019.8787797.

48. Somok, M. and Drew, A.H., A 13.9-nA ECG Amplifier Achieving 0.86/0.99 NEF/PEF Using AC-Coupled OTA-Stacking, 55, 414–425, 2020, article in press. doi: 10.1109/JSSC.2019.2957193.

49. Kosari, A., Breiholz, J., Liu, N., Calhoun, B.H., Wentzloff, D.D., A 0.5 V 68 nW ECG Monitoring analog front-end for arrhythmia diagnosis. *J. Low Power Electron. Appl.*, 1–13, 8, Jun. 2018. doi:10.3390/jlpea8030027.

3

Detecting COVID-19 Through Lung X-Ray Imaging: An Alternative Using Modified CNN Architecture

Ahan Chatterjee[1] and Sovik Mukherjee[2*]

[1]*Department of Computer Science and Engineering, The Neotia University, Sarisha, India*
[2]*Faculty of Commerce and Management, St. Xavier's University, Kolkata, India*

Abstract

The arrival of COVID-19 took the very existence of human race for a toss. In countries like India, where the majority of the population is concentrated in the rural areas and are subject to an affordability and infrastructural constraint, cannot afford sophisticated COVID-19 tests. But, X-Ray is widely obtainable across both the rural and urban belts of our country and comes at an affordable cost, even free at the government hospitals.

In the present research paper, we put forward a fusion-based DCGAN and CNN based neural net architecture which will generate synthetic COVID-19 infected lung X-Ray images from our fed data. Here we consider mainly two (2) output classes namely, malignant and benign. The novelty in this paper is that from the original X-Ray Image our model will generate a "predicted" image instantaneously using the DCGAN structure to understand the process of mutation. Also, the model predicts the class of the newly generated "predicted" image, whether it is COVID-19 positive or negative through the proposed CNN architecture. However, the paper that the success of deploying our model depends on the availability of the 5G network as the "predicted" X-Ray image along with the original X-Ray image of a patient needs to be transmitted to a central server from where it needs to be analyzed for further course of treatment as already specified. We have made an attempt to achieve the state of the art accuracy in our CNN model.

Corresponding author: sovik1992@gmail.com

Arun Kumar, Manoj Gupta, Mahmoud A. Albreem, Dac-Binh Ha and Mohit Kumar Sharma (eds.)
Wearable and Neuronic Antennas for Medical and Wireless Applications, (37–56) © 2022
Scrivener Publishing LLC

Keywords: CNN, deep learning, GAN, DCGAN, COVID-19, neural network, SARS-2

3.1   Introduction

The arrival of COVID-19 took the existence of human race at a toss. The number of patients being affected is increasing but the daily rate of being affected has however slowed down. Early detection of the disease is the need of the hour. There are various COVID-19 tests which are there in the market for confirmation, but most of them give false positive cases; only RT-PCR test gives the most accurate results. In countries like India, where the majority of the population is concentrated in the rural areas and are subject to an affordability constraint, cannot afford the test. Moreover, from our experience we have seen that such sophisticated tests for COVID-19 detection are not conducted outside the urban metros due to lack of infrastructural support. Thus, the concept of simplified method of testing comes into the picture—X-ray is widely available across both the rural and urban belts of our country and comes at an affordable cost, even free at the government hospitals. In this context, the novelty of our paper comes; we propose a CNN based architecture for detection of COVID-19 through X-ray images, and we deploy this model into a cloud based system through edge computing method. However, given India's three tier health system [in terms of Sub-centers and Primary Health Centers (PHCs) at the primary level, Community Health Centers (CHCs) at the secondary level and hospitals at the tertiary level] and the infrastructural constraints and lack of a specialized set-up at the primary and the secondary levels, for the working of our model we propose the creation of a central server (at any level having the required specialized doctors to address the concern of the COVID-19 affected patients, if infrastructure permits) from where the COVID-19 results generated based on the X-ray images can be analyzed and transmitted back to the sub-servers at different levels specifying the course of treatment then and there. This system of using X-ray images serves as an alternative to the sophisticated means of testing which requires transmission of the collected swab/blood samples physically. However, the success of deploying our model depends on the availability of the 5G network as the "predicted" X-ray image along with the original X-ray image of a patient needs to be transmitted to a central server from where it needs to be analyzed for further course of treatment as already specified. This greatly solves the problem of physical transmission and time to address those patients who are not in a position to access the facilities and are far off.

One of the most promising approaches of image synthesis is the concept of Generative Adversarial Networks (GAN). Data Augmentation becomes easier with the applicability of GAN as its capability of creating high quality realistic image from the training dataset. Along, with the data generation we need to classify those generated images into malignant and benign class for prognosis, and for this we use Convolutional Neural Network (CNN). CNN is highly capable to classify image based on the patterns and features present in an image. In this paper, our study focuses on generating new COVID-19 cells using GAN, which doesn't exist currently but it may appear in human body as a malignant cell due to constant mutation of COVID-19 cells. As the COVID-19 cells are constantly mutating it's becoming more dangerous than before, thus detection at an early stage becomes more important. Rapid detection could be done by analyzing the X-ray images of COVID-19 patients. In the present research paper, we put forward a fusion-based DCGAN and CNN based neural net architecture which will generate synthetic COVID-19 infected lung X-ray images from our fed data. The images which will be generated don't exist but can be formed in the near future due to the constant mutation of the virus. Afterwards, the synthetic image is passed through a CNN net architecture which will predict the output class of the synthetic image. Here we consider mainly two (2) output classes namely, malignant and benign. The novelty in this paper is that from the original X-ray image our model will generate a "predicted" image instantaneously using the DCGAN structure to understand the process of mutation (the "predicted" image does not exist at present and should have come up over a period of time through the mutation of the virus but here we do it instantaneously ensuring both cost effectiveness and detection of COVID-19 at an early stage). Also, the model predicts the class of the newly generated "predicted" image, whether it is COVID-19 positive or negative through the proposed CNN architecture. In this way we create a data library for new COVID-19 cells and classify them into classes for future prognosis.

The paper is structured as: Section 3.2 contains literature review, Section 3.3 contains our proposed methodology and model architecture, Section 3.4 contains results, and analysis on that backdrop, and Section 3.5 as concluding remark with future scope of study.

3.2 Literature Review

Shin *et al.* [1] used Image to Image Condition in GAN to generate the synthesized data to classify T1 brain tumor class on ADNI dataset.

They showed it can increase accuracy of classifier if trained in GAN images rather than original dataset. Iqbal *et al.* [2] proposes an innovative method of medical imaging using GAN (MI-GAN) to generate retinal images. The results show the newly formed image contains the structure from original image. Senaras *et al.* [3] proposed a conditional GAN technique (cGAN) to synthesize. Another work by Mahapatra *et al.* [4] shows to generate a high resolution image from a low resolution image through proposed model of P-GAN. Another work suggested implementing GAN for hyper spectral images, authored by Zhu *et al.* [5]. Fosto Kamga guy *et al.* proposed to use transfer learning algorithms such as VGG16, Alexnet, and InceptionV3 to create a fusion-schema model to generate images. Shang *et al.* [6] computed the missing data problem in a dataset using Cyclic-GAN technique. Gurumurthy *et al.* [7] proposed a DeLiGAN method to solve problems arise due to limited dataset for training. Premchand and Dutt [8] proposed a methodology through which GAN can be implemented for speech denoising. One of the major challenges in implementing CNN model is the requirement of cleaned required data, thus to solve synthetic images are taken [9].

3.3   Proposed Methodology

In current date also there is an acute shortage of medical image data through which we can carry out proper classification of COVID-19 Cells. In this section we propose a CNN based architecture which can classify synthetic image which has been formed using GAN architecture. The primary research findings of our proposed algorithm are as follows:

(i)   Generation of synthetic Lung X-Ray Image using proposed DCGAN Architecture

(ii)   Synthesize and validate the missing modalities in the synthetic generated images.

(iii)   Classifying the synthetic images using CNN Architecture.

3.3.1   Generative Adversarial Network (GAN)

The introduction of Deep Neural Network led to a paradigm shift in the applicability of AI in various fields. There is a massive advancement in the algorithm design due to increased accuracy of neural networks. Generative Adversarial Network is such one classic example of this. This algorithm is capable of generating synthetic images which simply doesn't exist but looks totally realistic. The GAN is based on the Game Theory approach where there are two players here, two neural networks which will try to optimize each other's result until the synthetic images has similar features with the original training data. GAN works on Zero-Sum principle, and has two blocks to generate the synthetic image namely generator and discriminator.

- *Generator:* Generator or the Generative neural network is mainly responsible for creating synthetic image with the goal to get undetected. It generates the image without training the features of the image of the input dataset, i.e. without learning semantics of the input image data.
- *Discriminator:* The discriminator neural network, learns to classify that the given sample is from the same data distribution or not. The major goal of a discriminator network is to detect the fake content in the set. It's basically a classifier network which classifies whether the image is real or not.

The GAN model is based on two separate convo neural net architecture. In our proposed architecture we have used Deep Convolutional GAN (DCGAN). The deep generative models are equipped to act against the backdrop of the difficulty in approximated computing, which is generated in Maximum Likelihood estimation and the variables used in the model are reported in Table 3.1. Thus training both neural networks gives us an upper hand to generate better results. The basic architecture of GAN is showed in Figure 3.1.

The maximization of D is represented in Eq. (3.1). The function should be maximized as it will give decision over real data.

$$E_{x \sim P_r(x)}[\log D(x)] \tag{3.1}$$

Table 3.1 Denotes meaning of variable. Source: Created by the authors.

Symbol	Meaning
P_z	Data distribution over noise z
P_g	The generator distribution over data x
P_r	Data distribution over real sample x
G	Generator
D	Discriminator

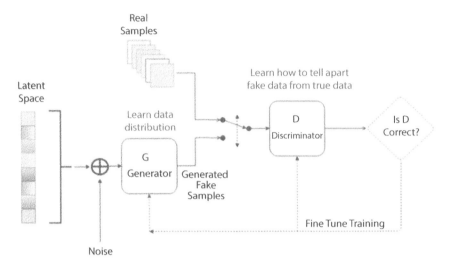

Figure 3.1 Architecture of GAN.

The fake sample function of $G(z)$ is represented in Eq. (3.2).

$$G(z) = z \sim p_z(z) \qquad (3.2)$$

D will return an output with a probability $D(G(z))$ of close to 0 in order to maximize the function given in Eq. (3.3).

$$E_{z \sim p_z(z)}[\log(1 - D(G(z)))] \qquad (3.3)$$

Similarly on the other hand we have architecture of G where the function should be minimized in order to get the subsequent result. The minimization formula is represented in Eq. (3.4).

$$E_{z \sim p_z(z)}[\log(1 - D(G(z)))] \qquad (3.4)$$

Thus combining the both parameters together e design the GAN network architecture as whole. The both parameters play a minmax game in which we have to optimize the function to get optimal result shown in Eq. (3.5)

$$min_G max_D L(D,G) = E_{x \sim P_r(x)}[\log D(x)] + E_{z \sim p_z(z)}[\log(1 - D(G(z)))] \qquad (3.5)$$

Solving the equation:

$$min_G max_D L(D,G) = E_{x \sim P_r(x)}[\log D(x)] + E_{x \sim P_g(x)}[\log(1 - D(G(x)))] \tag{3.6}$$

The term $E_{x \sim P_r(x)}[\log D(x)]$ doesn't have any impact on the result of G during gradient descent.

The loss function should be optimized in order to get the best results. Thus optimizing the loss function for D.

$$L(G,D) = \int_x (P_r(x)\log(D(x)) + P_g(x)\log(1 - D(x)))dx \tag{3.7}$$

We aim to optimize the value of $D(x)$ thus maximizing $L(G, D)$

$$\hat{x} = D(x), \qquad A = P_r(x), \quad B = P_g(x)$$

Substituting and calculating the values in the integral

$$f(\hat{x}) = A\log\hat{x} + B\log(1 - \hat{x}) \tag{3.8}$$

$$\frac{df(\hat{x})}{d\hat{x}} = A\frac{1}{\ln 10} - B\frac{1}{\ln 10}\frac{1}{1 - \hat{x}} \tag{3.9}$$

$$\frac{df(\hat{x})}{d\hat{x}} = \frac{1}{\ln 10}\left(\frac{A}{\hat{x}} - \frac{B}{1 - \hat{x}}\right) \tag{3.10}$$

$$\frac{df(\hat{x})}{d\hat{x}} = \frac{A - (A + B)\hat{x}}{\hat{x}(1 - \hat{x})} \tag{3.11}$$

Now, equating $\dfrac{df(\hat{x})}{d\hat{x}} = 0$ we will get the optimized value for D

$$D_o(x) = \frac{P_r(x)}{P_r(x) + P_g(x)} \in [0,1] \tag{3.12}$$

When both of our architecture are optimized we reach a state where $P_r = P_g$ and $D_o(x) = 1/2$

The new loss function is defined in Eq. (3.15).

$$L(G, D_o(x)) = \int_x^{\cdot} (P_r(x)\log(D_o(x)) + P_g(x)\log(1 - D_o(x))) dx \quad (3.13)$$

$$L(G, D_o(x)) = \log\frac{1}{2}\int_x^{\cdot} P_r(x) dx + \log\frac{1}{2}\int_x^{\cdot} P_g(x) dx \quad (3.14)$$

$$L(G, D_o(x)) = -2\log 2 \quad (3.15)$$

This is the optimized function of our GAN model, the model is arranged with Deep Layers which has been elaborately discussed in Section 3.2, the model architecture which we have implemented along with the our proposed algorithm.

3.3.2 Convolutional Neural Network (CNN)

The Convolutional Neural Network or CNN is a neural network architecture generally used to classify images. In this paper we have used CNN architecture after the last layer of our DCGAN model to classify the synthetic images which have been created by the generator.

CNN is often regarded as ConvNet, is equipped with deep feed forward architecture in it. Using that feature it makes it more able detect better than other fully connected layers. It is based on the concept of weight sharing. One of the major benefit in CNN is it takes very less amount of parameters thus it is safe from over fitting of the model.

3.3.2.1 General Model of CNN

In general ANN model takes one single input and output layer along with multiple hidden layers which act as the processing unit for the computation of the result. The vector x produces an output layer of Y, performing any particular function as represented in Eq. (3.16).

$$F(x, W) = Y \quad (3.16)$$

Here W denotes the weight vector which represents the interconnectivity strength of adjacent layer neurons. The general model of CNN consists of four layers namely, (a) Convolutional Layer, (b) Pooling Layer, (c) Activation Layer, and (d) Fully Connected Layer.

3.3.2.2 Convolutional Network

An image is constituted of pixels. The image is being classified by the feature extracted from the input layer, and based on the feature set the output class is being determined (see Figure 3.2). The adjacent layer neurons are connected among each other and they extract features from the receptive field to form the weight vector.

The weight vector is alternatively known as filter, or kernel, which slides over the input vector field to form the filter map as shown in Figure 3.3. This method of sliding over the input vector either horizontally or vertically is called Convolutional operation. This creates N feature map with the

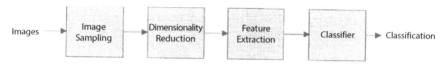

Figure 3.2 Basic constituent layer of CNN.

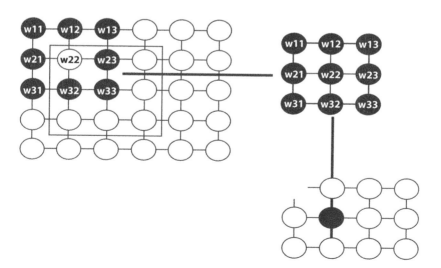

Figure 3.3 Weight vector initialization.

N features which have been extracted. The output which will subsequently act as input for the next layer is given in Eq. (3.17).

$$a_{ij} = \sigma((Wx_{ij} + b)) \qquad (3.17)$$

Where,
  x is the input vector
  W is the kernel which slides over an image
  b is the bias
  σ is the non-linearity introduced in the network architecture.

3.3.2.3   Pooling Layer

The primary aim to introduce pooling layer is to reduce the number of trainable parameters by introducing translation invariance. To perform this operation a window is being created and the input elements are passed through it a pooling function as shown in Figure 3.4. The maximum figure is taken from a matrix thus this method is max-pooling and the layer doing this is known as max-pooling layer.

3.3.2.4   Fully Connected Layer

The fully connected layer is being fed by the output which has been generated from the last two layers. The loss function, gradient descent reduces the cost function here over the training dataset, with constantly updating the weights in the layer with every passing epochs, where an epoch is defined as the journey of traversing the whole network.

3.3.2.5   Activation Function

The introduction on non-linearity is needed in the architecture, and in order to introduce that particular non linearity we generally use Rectified

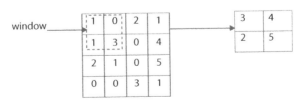

Figure 3.4 Pooling window.

Linear Unit (ReLu) activation function. It's chosen based on the advantages it gives like, doesn't allows the gradient to disappear which gives a major upper hand in loss calculation in CNN architecture. Sometimes we opt for leaky ReLu if a large gradient is being passing through the network, as at that time ReLu fades away.

3.3.2.6 Calculation of Gradient Descent in CNN Architecture

In the time of training through the filters, we use a technique named back-propagation to change and update the pre-initialized weights that are being allotted. The overall network is at first feed forwarded, and then the computing at every layer starts and total error component is introduced in the last layer. In order to compute an optimized network the computed gradients are being backpropagated the calculation is as follows.

After passing the input vector through Eq. (3.18) the next steps are being calculated.

$$C_q^l = \left(\sum_{p=1}^{n} S_p^{l-1} * K_{p,q}^l + b_q^l \right) \tag{3.18}$$

$$C_q^l = \left(\sum_{p=1}^{n} \sum_{u=-x}^{x} \sum_{v=-x}^{x} S_p^{l-1}(i-u, j-v) . K_{p,q}^l(u,v) + b_q^l \right) \tag{3.19}$$

Where,

n = Number of feature map in last layer

p, q = Layer map of last layer and previous layer

φ = Activation Function (ReLu or Sigmoid)

$*$ = Convolutional Operation

b = bias

x = Size of Filter

S_p^0 = Input Image where first ConvNet will be performed

Z = Pool Size Window

After Conv. Layer pooling is applied as represented in Eq. (3.20)

$$S_q^l(i,j) = \frac{1}{4} \sum_{u=0}^{z} \sum_{v=0}^{z} C_q^l (2i-u, 2j-v) \qquad (3.20)$$

Afterwards it's being passed through fully connected layers for output class prediction.

$$\widehat{output} = \sigma (w \times f + b)$$

Where, f is the final output vector

Then the result is passed through the softmax activation function, which is represented in Eq. (3.22).

$$y'(i) = \frac{e^{\widehat{output}}}{\sum_1^{labels} e^{\widehat{output}}} \qquad (3.22)$$

The loss function is computed through Eq. (3.23)

$$L = \frac{1}{2} \sum_{i=1}^{no.\,of\,training\,pattern} (y'(i) - y(i))^2 \qquad (3.23)$$

Where,

$$y'(i) = \textit{Target Output}$$

$$y(i) = \textit{Predicted Output}$$

The Error is now needed to be back-propagated

$$\Delta W(i,j) = \frac{\partial L}{\partial W(i,j)} \qquad (3.24)$$

$$\Delta W(i,j) = \frac{\partial L}{\partial wy'} \cdot \frac{\partial y'}{\partial W(i,j)} \qquad (3.25)$$

$$\Delta W(i,j) = \partial \left(\frac{1}{2} \sum_{i=2}^{P*} \frac{\left((y'(i) - y(i))^2 \right)}{\partial y'} \cdot \frac{\partial y'}{\partial W(i,j)} \right) \tag{3.26}$$

$$\Delta W(i,j) = (y'(i) - y(i)) \times \frac{\partial}{\partial W(i,j)} (\sigma(\sum_{j=1}^{no.\,of\,codes} W(i,j) \times f(j) + b(i))) \tag{3.27}$$

$$\Delta W(i,j) = \Delta y'(i) \times f(j) \tag{3.28}$$

$$\Delta b = \frac{\partial L}{\partial b(i)} \tag{3.29}$$

3.3.3 Proposed Algorithm

In this section we propose our novel algorithm through which we can generate and then classify the blood COVID-19 cells.

Input:

a. *Discriminator: Original Datset*

b. *Generator: Noise with Partial Range of Original Datset*

Output: Predicted Class (Malignant or Benign) of Synthetic Blood Cancer Cell Image

Start

G:

$Sy_{Smpl} = Gn\,(GUns + Data\,(Lb, UB))$

Do

{

D:

Diff (Sy_{Smpl})

D *Diff* N_{smpl} & Sy_{Smpl} *using MLE*

if (*MLE = High*)

Accept Sy_{Smpl}

Else

G (*Fdbk*)

G:

$Sy_{Smpl} = Gn\,(GUns + Data\,(Lb, UB)) + Fdbk$

$$G \; clts \; Error \; between \; Sy_{Smpl} \; \& \; N_{Smpl}$$
} *while* (Sy_{Smpl} = *Accepted*)
$$CNet(Sy_{Smpl}):$$
$$for \; (i = 0:3)$$
{
$$ConvoNN(St = 16 \; (Sy_{Smpl})) \; \& \; fv \; Generate$$
$$Mplng \; (fv)$$
$$Dpl \; (0.2)$$
$$St + +;$$
}

$$FlCl \; (Mplng)$$
$$Opl \; (FlCl)$$
End

The meaning of the variables used in the proposed algorithm is explained in Table 3.2.

Table 3.2 Variable and meaning in the proposed algorithm. Source: Created by author.

Variable	Meaning
Sy_{Smpl}	Synthetic Sample
Gn	Generate
GUns	Gaussian Noise
Lb, UB	Lower Bound, Upper Bound
Diff	Differentiate
N_{smpl}	New Sample
MLE	Maximum Likelihood Function
Fdbk	Feedback
clts	Calculates
CNet	CNN
ConvoNN	Convolutional Network

(Continued)

Table 3.2 Variable and meaning in the proposed algorithm. Source: Created by author. (*Continued*)

Variable	Meaning
fv	Feature vector
Mplng	Max Pooling Layer
Dpl	Dropout Layer
St	Stride
FlCl	Fully Connected Layer
Opl	Output Layer

3.3.4   Model Architecture

The classification model of CNN is based after the GAN where the synthetic images of lung image are being generated. The generator network takes an input vector of size 100, of which are taken as sample at random from a normal distribution. The input is being propagated through the fully connected internal layer architecture and as a result it creates a RGB image with constituent's 128 * 128 * 3 pixels.

The model is equipped with a fully connected layer along with 4 deconvolutional with stride 1 and a kernel size of 4 * 4. All the layers except the output layers uses Leaky ReLu as its activation function, and the last output layer uses tanh activation function.

The discriminator is designed with a classic architecture. The input network is a n RGB image with the pixel size of 128 * 128 * 3. Similar architecture has been followed here with four Convolutional networks and keeping other layers same as the generator architecture. Except the output layer we have used Leaky ReLu activation function with a slope of 0.1, and the last layer is fitted with sigmoid function which will eventually return the probability of the image being fake or real.

After passing through the DCGAN architecture the image is fed into a Convolutional network which will act as a classifier to classify whether the generated images are good enough to get detected as COVID positive or negative cell through an automated process (see Figures 3.5 and 3.6). In this architecture, we have used four Convolutional layers, along with a dropout layer with 0.2 probabilities to minimize the over fitting in the model.

The basic architecture stands as then, the dataset is passed through the generator network architecture, it creates synthetic image as its output.

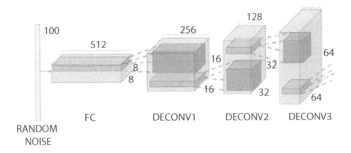

Figure 3.5 Discriminator architecture.

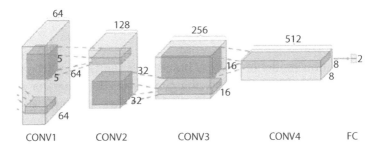

Figure 3.6 Generator architecture.

Then that image is passes through the discriminator network where it's been detected for being original or fake. Once the image is being approved then it's passed through our Convolutional network architecture. In that step it's being predicted whether the synthetic image which have been generated from our DCGAN architecture.

3.4   Results

In this section we discuss the results which have generated through our model.

At first, we show the images which are being generated from our DCGAN Architecture.

Table 3.3 clearly shows that our DCGAN model is creating images which are synthetic, but are good enough to pass the automated system of our CNN architecture. The training set contains around 10,000 lung X-Ray Image.

Table 3.4 shows the accuracy of our CNN architecture model.

Table 3.3 Shows image formed after epochs through our DCGAN architecture. Source: created by Authors.

Parameter value	Image set
Image Size = 128 * 128 Noise Size = 10 Discriminator Learning Rate = 0.00034 Generator Learning Rate = 0.00036 After 50 Epochs.	
After 70 Epochs with constant parameter	
After 120 Epochs with constant parameter	
After 200 Epochs with constant parameter	
After 250 Epochs with constant parameter	
After 300 Epochs with constant parameter	

We have created our CNN architecture based on 10,000 training images which are different from the images which have been passed through the DCGAN architecture. The validation set consists of around 1,000 images along with 3,000 images in test set (see Table 3.4, Figures 3.7 and 3.8).

Table 3.4 Accuracy of CNN architecture. Source: Created by author.

Model	Accuracy
Validation Accuracy	94.56%
Training Accuracy	92.32%

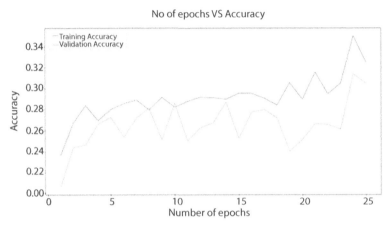

Figure 3.7 Accuracy curve for CNN model. Source: Created by authors.

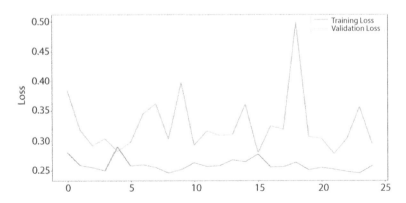

Figure 3.8 Loss curve of CNN architecture. Source: Created by authors.

3.5   Conclusion

The paradigm shift in the field of 5G communication has enabled user mobile connectivity to the next level. 5G connectivity aiming to provide a 1,000 times higher mobile data per area, with a 100 times higher user data rate, with a significant drop of latency. The major novelty in the development of 5G connectivity in remote healthcare system is to enable the Edge Computing Scenario across the nation. In the uncertain time of COVID-19 the data collection of patients and uploading it through cloud computing architecture to make the necessary information available is the need of an hour. Based on that information various government policies have been taken, along with that spread, and mutation scenarios have also been analyzed. Such high amount of data transfer requires high latency and 5G model can handle such things with much more ease.

In this paper, we have proposed a DCGAN based architecture which can generate synthetic COVID-19 infected lung X-ray image using the two neural net architecture of it, namely the generator, and discriminator. Later, we have passed that newly generated image through a CNN architecture, which is trained to predict the output classes of the COVID-19 cell images.

We have trained our DCGAN architecture on 10,000 images, and our CNN architecture is being trained on another different set of 12,000 blood cell images. Both the classes malignant or benign were present in that dataset. The validation set consisted around 1,000 images along with 3,000 test set images to test the predictability of our model.

We have achieved a high accuracy of 92.32% on the validation set, and through which we get high probability of getting correct class in the output even through synthetic image.

References

1. Karako, L., Nygate, Y., Roitshtain, D., Rubin, M., Stein, O., Shaked, N.T., Turko, N.A., TOP-GAN: Stain-free COVID-19 Cells cell classification using deep learning with a small training set. *Med. Image Anal.*, 57, 176–185, 2019.
2. Cao, Q., Gan, Z., Hao, L., Mao, Z.W., Yang, J., Zhou, D., Zhao, J.X., Simultaneously inducing and tracking COVID-19 Cells cell metabolism repression by mitochondria-immobilized rhenium (I) complex. *ACS Appl. Mater. Interfaces*, 9, 16, 13900–13912, 2017.
3. Kong, X., Su, H., Yang, L., Xie, Y., Xing, F., Beyond classification: Structured regression for robust cell detection using convolutional neural network, in: *International Conference on Medical Image Computing and Computer-Assisted Intervention*, pp. 358–365, Springer, Cham, 2015.
4. Chen, Y.W., Han, X.H., Ipponjima, S., Iwamoto, Y., Kitrungrotsakul, T., Yokota, H., Takemoto, S.,., A cascade of CNN and LSTM network with 3D anchors for mitotic cell detection in 4D microscopic image, in: *ICASSP 2019-2019 IEEE International Conference on Acoustics, Speech and Signal Processing (ICASSP)*, pp. 1239–1243, IEEE, 2019.

5. Huang, J., Pan, H., Xu, Z., An effective approach for robust lung COVID-19 Cells cell detection. *International Workshop on Patch-based Techniques in Medical Imaging*, pp. 87–94, Springer, Cham, 2015.

6. Chefd'Hotel, C. and Chen, T., Deep learning based automatic immune cell detection for immunohistochemistry images, in: *International Workshop on Machine Learning in Medical Imaging*, pp. 17–24, Springer, Cham, 2014.

7. Chen, S., Guan, Q., Hu, H., Huang, Y., Zhang, J., COVID-19 Cells detection in phase-contrast microscopy images based on Faster R-CNN, in: *2016 9th International Symposium on Computational Intelligence and Design (ISCID)*, vol. 1, pp. 363–367, IEEE, 2016.

8. Asopa, P., Goswami, A.K., Indolia, S., Mishra, S.P., Conceptual understanding of convolutional neural network—A deep learning approach. *Proc. Comput. Sci.*, 132, 679–688, 2018.

4

Wireless Body Area Network Antenna

Inderpreet Kaur, Hari Kumar Singh* and Tejasvini Thakral

Department of Electronics and Communication Engineering,
M.J.P. Rohilkhand University, Bareilly, India

Abstract

The increase in average lifespan and health cost in many developed nations are catalysts to innovation in health care. These factors along with the advances in miniaturization of electronic devices, sensing, battery and wireless communication technologies have led to the development of Wireless Body Area Networks (WBANs). WBANs consist of smart miniaturized devices (motes) that are able to sense, process and communicate. They are designed such that they can be worn or implanted, and monitor physiological signals and transmit these to specialized medical servers without much interference to the daily routine of the patient.

A rectangular microstrip antenna with surface integrated waveguide (SIW) and defected ground structure (DGS) is proposed for WBAN application. SIW is formed using metallic via array with specific location and dimension. Microstrip line feed excites rectangular patch. The proposed work has optimized the radius of via, length of feed, dimension of DGS. The gain of the proposed configuration is 1.9dBi at 2.45 GHz. The 3D polar plot shows the omnidirectional pattern.

The challenges faced by WBAN are energy, mobility, security and communication. WBAN devices are generally wireless, therefore they are portable and roam free. So the power to the device of the network is provided with the help of batteries, since the sensor that implanted in the body are so small that battery cannot sustain more than a month, Energy harvesting technique is one solution to the power issue. WBAN security is more critical orbit as it has been connected to the physical system. Security required in internal communication are data authenticity done by public key cryptography, encryption is done to achieve

*Corresponding author: harsdik@gmail.com

Arun Kumar, Manoj Gupta, Mahmoud A. Albreem, Dac-Binh Ha and Mohit Kumar Sharma (eds.) Wearable and Neuronic Antennas for Medical and Wireless Applications, (57–84) © 2022 Scrivener Publishing LLC

data confidentiality. Mobility can pose serious problems in some applications like e-health care even posture do effect the communication.

Keywords: cryptography, wearable monitoring system, cardiovascular diseases, blood oximetry, switching rate, frequency drift, enabling priority and gain deteriorates.

4.1   Introduction

The desk bound and inactive lifestyle is increasing the prevalence of persistent diseases. Wireless body area networking (WBAN) technology has the potential to provide an unparalleled opportunity for real-time healthcare and fitness monitoring in ambulances, emergency rooms, operation theaters, recovery rooms, clinics, homes and even on the move, so that most diseases could be prevented through early detection and doctors could give patients efficient advice to improve their health [4, 5].

One form of the Wearable Physiological Monitoring System (WPMS) is the Wearable Body Area Networks (WBAN) which monitors the health status of the wearer for long durations. The chapter discusses a WBAN based wearable physiological monitoring system to monitor physiological parameters such as Electrocardiogram (ECG) and Electroencephalogram (EEG) acquired using a textile electrode, Photoplethysmogram (PPG), Galvanic Skin Response (GSR), Blood Pressure derived from analysis of Pulse Transmit Time (PTT) and body temperature [3].

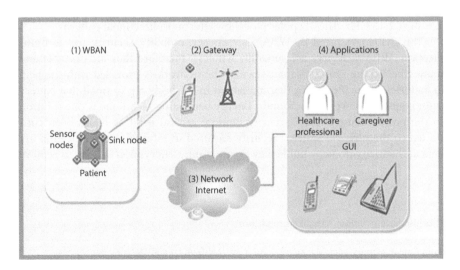

Figure 4.1 The block diagram of WBAN.

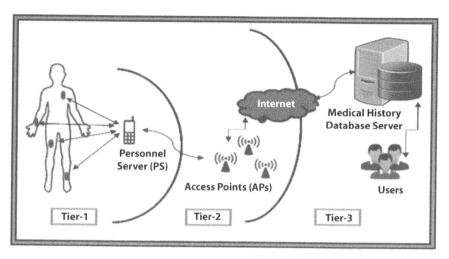

Figure 4.2 The three tiers involved in WBAN and its medical needs. Source: https://doi. org/10.1016/j.eij.2016.11.001.

The WBAN consists of three sensor nodes that are placed strategically to acquire the physiological signals and the sensor nodes communicate to a chest/wrist worn sink node also known as wearable data acquisition hardware. The sink node receives physiological data from the sensor nodes and is transmitted to a remote monitoring station as shown in Figure 4.1. The remote monitoring station receives the raw data and it is processed to remove noises, such as power line interference, baseline wander and tremor in the signals and the information is extracted and displayed. The WBANs are designed using the ZigBee wireless communication modules to transmit and receive the data in Figure 4.2. At the remote monitoring station the physiological parameters such as heart rate, pulse rate, systolic, diastolic blood pressure, GSR and body temperature are continuously monitored from the wearer. The data acquired from the wearable monitoring system is validated qualified medical device [6, 7].

The use of WBANs for medical applications is aimed at the continuous monitoring of an individual's physiological attributes, such as blood pressure, heart beat and body temperature. Medical applications of WBANs can be further classified into two subcategories which are as follows:

4.1.1   On Body WBANs

Wearable medical applications of WBANs not only bring convenience to patients in their daily life, but also increase the efficiency with which doctors can monitor and observe patient healthy in real time. Most sensors for

On-body WBANs operate in the Industrial, Scientific and Medical (ISM) band. Some of these applications are mentioned below. Traditional treatment of sleep disorders disturbs the patient's motion and initiates artefacts and noise that reduce the signal quality. However, WBANs are capable of delocalization of the intelligence and instruments in their sensor nodes and removal of all cable interconnects as shown in Figure 4.3. There are a number of wearable sensors and devices already available for use on the human body that are readily connectable to WBANs, including: accelerometers, blood glucose sensors, blood pressure monitors, ECG sensors, O_2 saturation monitors and thermometers. These promising applications are designed to benefit patients and doctors allowing continuous monitoring and mobility. Initial applications of BANs are expected to appear primarily in the healthcare domain, especially for continuous monitoring and logging vital parameters of patients suffering from chronic diseases such as diabetes, asthma and heart attack. A BAN in place on a patient can alert the hospital, even before they have a heart attack, through measuring changes in their vital signs. A BAN on a diabetic patient could auto inject insulin through a pump as soon as their insulin level declines [8, 9].

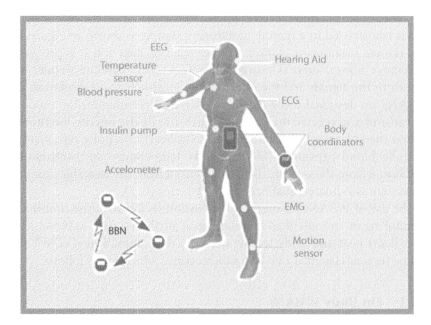

Figure 4.3 Patient monitoring using on body a wireless body area network (WBAN). Source: https://doi.org/10.3390/jsan8020027.

4.1.2   In Body WBANs

In-Body WBAN, another class of WBAN, is the implanted WBAN with frequency band of 402–405 MHz. This class of applications describes nodes implanted in the human body either underneath the skin or in the blood stream; some examples of systems that could utilize implanted WBAN nodes are described in Figure 4.4.

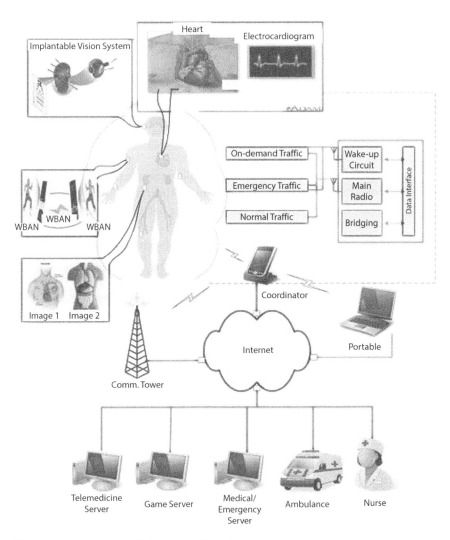

Figure 4.4 Application of In body WBAN in the various areas of medical and non-medical science. Source: DOI: 10.35940/ijitee.I1086.0789S19.

WBANs could provide frequent monitoring of blood sugar levels, used for example, in people suffering from diabetes through the implanting of sensors in the blood stream. This could help to reduce the risk of fainting and eliminating the risk of loss of blood circulation leading to blindness and other complications. Cardiovascular diseases could be significantly reduced or prevented, through monitoring heart rhythm and blood flow via sensors embedded in the body. The development of cancer detecting/ monitoring sensors as WBAN in body nodes will enable physicians to continually diagnose cancerous tumors without requiring observations providing more timely analysis and treatment as it includes the examination of gastrointestinal tract [6, 12].

4.1.3   Non-Medical Applications for WBAN

Non-Medical applications of WBANs can be further classified into six sub-categories as follows:

a) Entertainment and gaming Appliances such as microphones, MP3-players, cameras, head-mounted displays and advanced computer appliances can be used as devices integrated in WBANs. Body sensors enable game players to perform actual body movements, such as boxing and shooting, that can provide feedback to the corresponding gaming console, thereby enhancing their entertainment experiences.

b) Secure Authentication. This application refers to utilizing both physiological and behavioral biometrics such as iris recognition, fingerprints and facial patterns. This is one of the key applications of WBANs, where the issue of duplicability and forgery has motivated the use of new behavioral/ physical characteristics of the human body. The WBAN can identify the wearer based on the unique information from sensors around the body.

c) Personal information sharing Private or business information can be stored in body sensors and used to automate many daily life applications such as shopping and information exchange.

d) Emergency (non-medical) Off-body sensors (e.g. those built into a house) are capable of detecting non-medical emergencies such as home fires or the presence of flammable/ poisonous gases in the home; these sensors can communicate this information to body-worn devices, warning the wearer of the emergency condition. WBANs can also be used by policemen and fire-fighters in a similar manner.

e) Sports and Fitness Significant parameters like blood pressure, heart beat, blood oximetry and posture can be attained and collated through a single WBAN during an athletes' training. It is possible for WBANs to provide motion capture aiding athletes to improve their performance and prevent injury. They can also provide useful information relating to training schedules.

f) Military and Defense WBANs enable a soldiers activities to be monitored as well as providing the soldier with information about the surrounding environment in real time (for example during a battle), which results in better battlefield management and the ability to better avoid threats.

4.1.4   Principle of Operation

The working principle of wireless body area networks is depicted in Figure 4.5. Block diagram Sensor detects the input signals based on physical change due to reaction. A transducer converts the input into electric signal and amplifies the electric signal to process it using the processor. The processed signal is displayed finally. Transducer is used to convert one form of energy into other such as movement, electrical signals, radiant energy, thermal or magnetic energy. Actuator is a component in the sensor which is used to control the whole system by a control signal. In the transmitter, carrier frequency is less and then the signal is modulated using high modulation techniques. In the receiver side, low noise amplifier is used to remove noise. The signal is demodulated using demodulator and filtered out to obtain to original signal. In wireless body area networks, sensor nodes are placed in or around the body of the person to capture signals which is used to measure parameters such as blood pressure, temperature, ECG, EEG and EMG. Sensor node should be placed inappropriate position [2, 10, 11].

Figure 4.5 Wireless body area networks.

4.1.4.1   Transmitter

The working principle of a Transmitter in wireless body area networks is depicted in Transmitter Source: A transmitter is a device which sends information or data in the form of electromagnetic waves over the wired or wireless channel to the desired destination (receiver). The transmitter operation involves encoding, modulation and amplification. Encoding involves conversion of data into bits. The transmitter with low power consumption supports WBAN requirements. However, they can only be applied for short range communication. Similarly, scientists have designed an analytic model for wave propagation in WBAN. They calculated path gain vs path length and location. The path gain depends on frequency and location. Few researchers used ZCD (Zero Correlation Duration) to avoid MAI (Multiple Access Interference). BER (Bite Error Rate) is improved in ZCD compared to Pseudo noise code (PN). Researches have investigated the characteristics of UWB and draw backs of UWB such as path loss, power delay and he tested the above in indoor environment. The research is useful in designing high speed UWB.

Few researchers have estimated channel between half wavelength dipoles. Cross layer design is used for consuming energy. Path loss and time domain characteristics of channel are also estimated. In another work researchers have controlled transmission power to increases the life time of sensor nodes and link reliability. It increases sensor transmitter receiver International Conference on Innovations in Power and Advanced Computing Technologies [i-PACT2017] 3 the life time by 9.86%. It reduces packet loss by 30.2%. The studies about modulation schemes and the channel are estimated for UWB impulses radio WBAN system. Path loss and delay spread are created due to waveform distortion. BER is analyzed based on waveform distortion. Few researchers controlled the transmission power using ATPC. Gain estimation algorithm is used for on body to on body propagation channel. From the RSSI, the output power of transmitter is known, and then beacon power is used to calculate the gain. Researchers have used differential modulation with varying frequency techniques for node operation. BFSK (Binary frequency shift Keying) is used for varying frequency reduces circuit complexity, BER, phase noise, interface effect and energy wastage. Similarly, researchers have used various numerical methods to measure UWB signal propagation and results are compared. This model can reduce computational effect in future. Few scientists introduced compressed sensing theory which is used to evaluate multipath fading channel. They have obtained 20% reduction for path loss, 10% bit error rate,

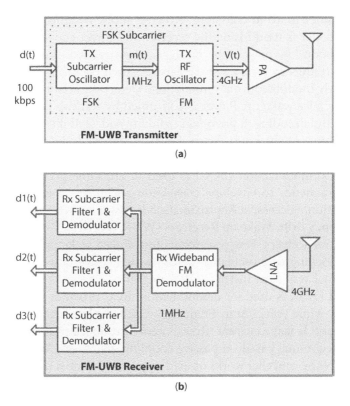

Figure 4.6 (a) Transmitter of wireless body area networks. (b) Receiver of wireless body area networks. Source: DOI: 10.3390/s141018009.

signal amplitude increased by 25% as the distance between transmitter and receiver increases.

Researchers introduced Carbon Nano Tubes (CNTs) which were used for data transmission lines. They have measured data collected, and its electrical properties. They are suitable for micro wave applications and easily accepted by biological tissues. However, scientist used a Switch combining with transmitter power control technique for controlling the transmitter power over a fixed transmitter and reducing switching rate. They have increased the network life time, radio communication and reliability as shown in Figure 4.6(a).

4.1.4.2   Receiver

A communication receiver is a device used to receive the information or data from the transmitter. Scientists have done modulation shift keying for transmitter to avoid emissions. Receiver block designed using CMOS analog decoder is used inside the receiver and direct conversion receiver used to obtain low power. The receiver design reduces the link loss, I/F noise and frequency errors. Few researchers used Gaussian on off keying/pulse width modulation for modulation. Detection schemes in receiver reduce the false wake up. Power is consumed by wakeup receiver, packet reception and decoding. It provides sensitivity and flexibility when receiver circuit changes as shown in Figure 4.6(b). Scientists designed a FSK wake up receiver using CMOS technology satisfied WBAN requirements. Body Channel Communication (BCC) is used to consume power; (IL-DCO) helps in frequency to envelope conversion and helps in amplifying and demodulation. Successive Approximation Register (SAR) is used to calculate frequency drift. Wake up Receivers (WURX) provides stable reference clock for transceivers. Few researchers introduced wake up receiver with injection-locking ring oscillator (ILRO) which converts weak signal into strong signal was used instead of RF amplifiers for power consumption. PLL based DFSK is used for demodulation. BCC provides sensitivity and selectivity, while AFC (Auto Frequency Oscillator) reduces temperature variation and leakage current. Researchers introduced mutual noise cancellation technique based on passive coupling which acts as a differential inductor. Noise canceller works under low power. Scientist used Maximum likelihood detectors for Pulse Position Modulation (PPM) and Transmitted Reference Pulse Amplitude Modulation (TR PAM) was designed [24, 25].

The performance of the receiver is increased and is suitable for non-coherent receivers. The parameters power delay profile, path loss, and multi path amplitudes are estimated. However authors introduced a Soft Decision Decoder (SDD) which provided a gain of 1 db when compared with hard decision decoder (HDD). It reduces test patterns with low voltage circuits. The performance of BER is improved, energy consumption 94%, SNR 5db is achieved. Few researchers used DFSK modulation technique for short range and low power communication. Circuit designed is reduced. BER and phase noise is estimated. Researchers used RF envelope detection for low power and short range communication. Direct modulation architecture and CMOS technology is used. System performance is improved System performance is improved by transmitter and receiver design. Few scientists investigated physical layer design of WBAN by using low cost base band transceiver. Digital timing synchronization is used for packet synchronization and data recovery. BER and packet error rate is estimated. Scientists have done Cooperative diversity for WBAN for monitoring a sleeping person. Two hop transmissions are used to increases the diversity gain [2, 6, 11].

4.1.5   Design Aspects of WBANs

Medical monitoring applications have specific hardware and network requirements to ensure their functions and to solve encountered problems.

Network requirements can be listed as follows:

1. Range: WBAN allow the sensors in on or around the same body to communicate with each other, so 2–5 m range is enough
2. Interference: Between the transmissions of different sensors from the same application and also from the different applications (because it would be possible to have a lot of sensors on the same body) and with other sources (people close to each other can also have their own WBAN), interference should be suppressed as much as possible to satisfy reliable wireless communication [1, 2].
3. Network density: With the diversification of WBAN applications, people should be able to have sensors for different applications and different body area networks on them. WBAN standard allows two to four networks per m².
4. Sensors number per network: Monitoring applications need a lot of sensors or actuators on the same body area network. WBAN standardization group expects a maximum of 256 devices per network.
5. Quick time of transmission: The goal of monitoring applications is a collection of information in the real time, so rapid transmission is necessary.
6. In-body environment: Wireless network technologies have always been confronted to the problem of obstacle between transmitter and receiver and the path loss in the propagation medium. The problem here is more important because most of the time, signal must propagate through human tissue.
7. Security/encryption: The transmitted data needs to be protected and integrity of received information should be provided.
8. Quality of Services (QoS) and reliability: They are crucial for the real time vital information and the alarm message transmission. WBAN standard has to provide error detection and methods of correction. The QoS needs to measure delay and delay variation, it must be flexible because each application has specific requirements, and must support real time transmission. Latency (time delay) must be inferior to 125 ms for the medical applications and to 250 ms for the

non-medical applications. Jitter (variation of latency) also must be controlled.

9. Enabling priority: WBAN has to support different type of traffics: periodical and burst traffic. Emergency traffic (alert in the case of heart attack or other serious problems) must have the priority on other messages on the network.

10. Support for different data rates: Applications have very different requirements of the data rate (from 10 kb/s to 10 Mb/s). Most of the time, medical applications need lower data rate whereas non-medical applications (particularly multimedia applications) need the most important data rate.

11. Compatibility with other PANs: WBAN should be able to communicate with devices using other PAN, such as Bluetooth.

4.1.6   Hardware Requirements

1. Ultra low power consumption: This depends on applications and on the nature of the sensors (in or on body). This constraint exists to allow sensors to have the smallest batteries and to avoid people who have to always change their batteries. An effective saving mode should be desirable

2. Parallel solutions can be mentioned: some sensors with very low power consumption can harvest the body power: temperature, vibration instead of batteries. Another one could be to charge batteries by induction.

3. Suitable sensors: Sensors should be small; they must not hinder mobility and users' life. Monitoring should be as transparent as much as possible for them.

4. Lifetime: Sensors must have a long lifetime, particularly in-body sensors.

5. Low complexity: They are associated with the low cost which need to be easily produced to have a low complexity. An easy way to implement the system is mass adoption of the system by everyone.

There are several advantages introduced by using wireless WBANs which include: Health Technology [20, 21].

1. Flexibility: Non-invasive sensors can be used to automatically monitor physiological readings, which can be forwarded to

nearby devices, such as a cell phone, a wrist watch, a headset, a PDA, a laptop, or a robot, based on the application needs

2. Effectiveness and efficiency: The signals that body sensors provide can be effectively processed to obtain reliable and accurate physiological estimations. In addition, their ultra-low power consumption makes their batteries long-lasting due to their ultra-low power consumption.

3. Cost-effective: With the increasing demand of body sensors in the consumer electronics market, more sensors will be mass-produced at relatively low cost, especially in gaming and medical environments [16, 22, 23].

4. Architecture of a Wireless Body Area Networks: A wireless body area network is composed by one or more Body Sensor Units (BSU), one Body Central Unit (BCU) and eventually a link with long range network. Different BSUs collect information about respiration rate, pulse, glucose rate, ECG, and other physiological signs. Each BSU transmits its data to the BCU. Sensors can have different behaviors to communicate with BCU. As the first option, they may send measured information continuously. This seems to be more intuitive but it is a power consuming method because transmitting signals using wireless communication results with high power consumption. Therefore, most of the time, this solution is not acceptable. Secondly, they send measured data periodically. This one allows real time monitoring which reduce power consumption for sensors. As another option, they can transmit a log file once a day. Physiological signs are measured periodically and values are recorded on a file that sent by sensors, for example, once a day. It allows monitoring while minimizing the number of sent messages and so energy consumed. Finally, they may send an alert message when measured physiological signal gets an abnormal value. The previous behaviors are for the normal cases, while this one is for emergencies. We do not want this message interferes with other ones so it needs more energy. The MAC layer also has to support priority message because of the time is precious in emergencies. The BCU analyzes all collected information, and acts in serious cases [13, 17, 19].

4.2   Literature Review

Design of a Dual-Band Antenna for WBAN Applications	The 2015 workshop on antenna technology	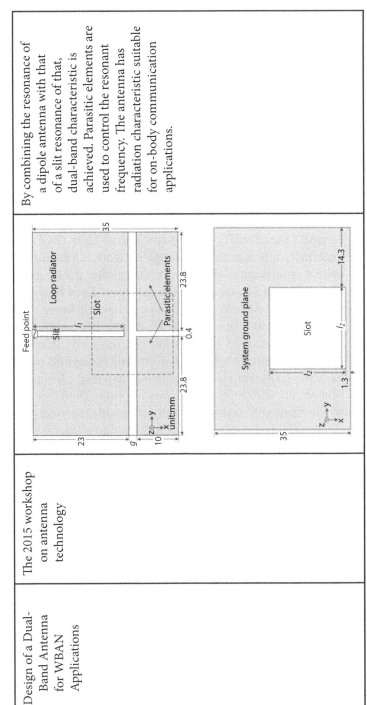	By combining the resonance of a dipole antenna with that of a slit resonance of that, dual-band characteristic is achieved. Parasitic elements are used to control the resonant frequency. The antenna has radiation characteristic suitable for on-body communication applications.

(Continued)

| Antennas for intraoral tongue d drive system at 2.4 GHz | IEEE Transactions on Microwave Theory and techniques | | The three types of antennas (patch, PIFA, and dipole) at 2.4 GHz to The results showed that the patch antenna has the highest gain in both closed- and open-mouth conditions, with −10.6 and −9 dB received power. The dipole antenna, on the other hand, demonstrated the widest bandwidth and best performance in the cross-polarization. |

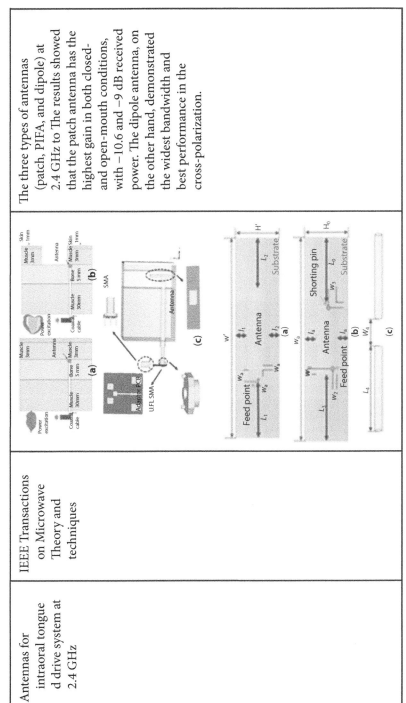

(Continued)

| An ultra-wide band low-SAR flexible metasurface enabled antenna for WBAN applications. | 3rd August, 2019 Springer - Verlag Germany, Part of Springer Nature, 2019 | 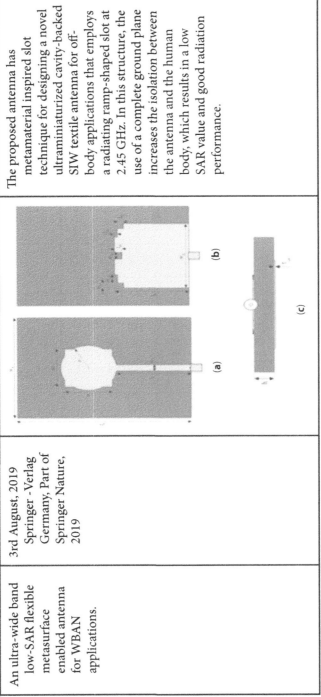 | The proposed antenna has metamaterial inspired slot technique for designing a novel ultraminiaturized cavity-backed SIW textile antenna for off-body applications that employs a radiating ramp-shaped slot at 2.45 GHz. In this structure, the use of a complete ground plane increases the isolation between the antenna and the human body, which results in a low SAR value and good radiation performance. |

(Continued)

| Design of WBAN For Biotelemetry operation. | Intelligent decision technologies | 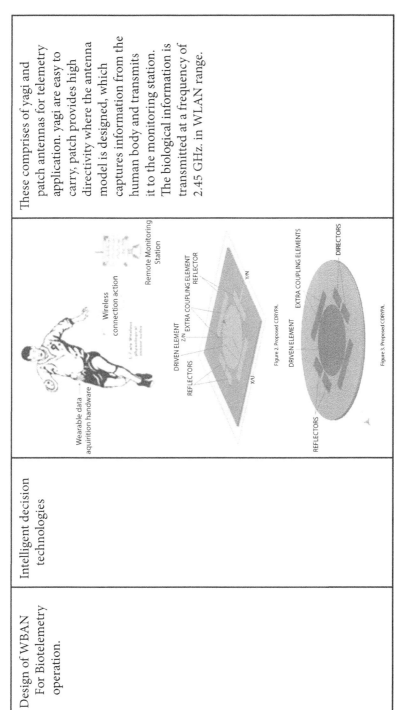 Figure 2. Proposed CDRYPA. Figure 3. Proposed CDRYPA. | These comprises of yagi and patch antennas for telemetry application. yagi are easy to carry, patch provides high directivity where the antenna model is designed, which captures information from the human body and transmits it to the monitoring station. The biological information is transmitted at a frequency of 2.45 GHz. in WLAN range. |

(Continued)

Ultraminiaturized Metamaterial inspired SIW Textile antenna for Off-Body Applications	IEEE Antenna and wireless propagation letters, VOL 16, 2017		The antenna was to operate at 2.45 GHz, the chosen substrate was wool felt with a height of 3 mm, dielectric permittivity of $\varepsilon r = 1.4$, and loss tangent $\tan\delta = 0.0254$. The electromagnetic properties of the wool were measured using the well-known Nicolson–Ross–Weir method.
Design of a V-Band Wideband circularly polarized Microstrip patch array elements for applications in body area network.	IEEE, 2017		A circularly-polarized, cavity-backed microstrip patch antenna, aperture-coupled fed through SIW, is presented. The antenna provides an impedance bandwidth of 13%, with a gain of 7.2 dB and cross-polarized levels of 18 dB at 60 GHz, and an axial ratio bandwidth of 7.6% using a single feed.

(Continued)

| An All-Textile SIW Cavity-Backed Circular Ring-Slot Antenna for WBAN Applications | IEEE Antennas and Wireless Propagation Letters, VOL. 15, 2016 | 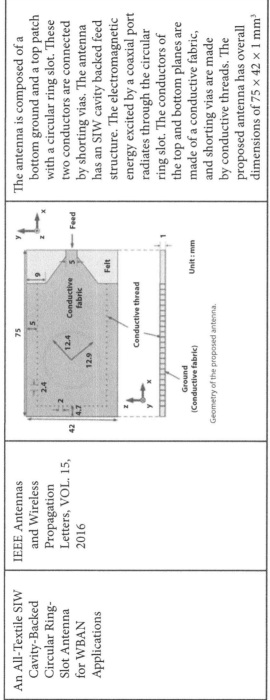 Geometry of the proposed antenna. | The antenna is composed of a bottom ground and a top patch with a circular ring slot. These two conductors are connected by shorting vias. The antenna has an SIW cavity backed feed structure. The electromagnetic energy excited by a coaxial port radiates through the circular ring slot. The conductors of the top and bottom planes are made of a conductive fabric, and shorting vias are made by conductive threads. The proposed antenna has overall dimensions of $75 \times 42 \times 1$ mm^3 |

(Continued)

Circular Ring-Slot Antenna Fed by SIW for WBAN Applications	7th European conference on antenna and propagation.	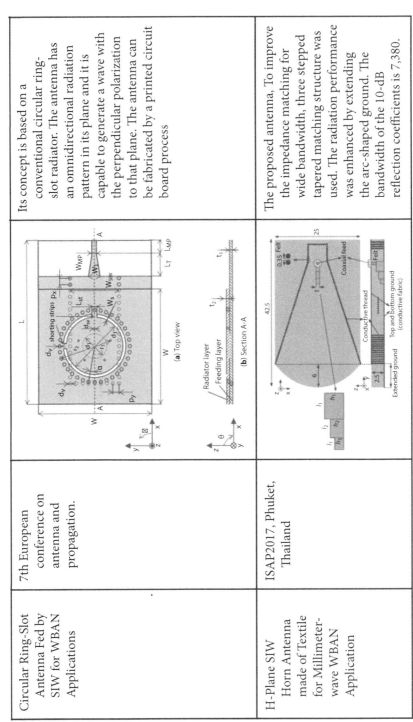	Its concept is based on a conventional circular ring-slot radiator. The antenna has an omnidirectional radiation pattern in its plane and it is capable to generate a wave with the perpendicular polarization to that plane. The antenna can be fabricated by a printed circuit board process
H-Plane SIW Horn Antenna made of Textile for Millimeter-wave WBAN Application	ISAP2017, Phuket, Thailand		The proposed antenna, To improve the impedance matching for wide bandwidth, three stepped tapered matching structure was used. The radiation performance was enhanced by extending the arc-shaped ground. The bandwidth of the 10-dB reflection coefficients is 7,380.

(Continued)

(Continued)

| On-Body Resilient SIW based Antenna for WBANs Applications | 2017 Twenty-third National Conference on Communications (NCC) | 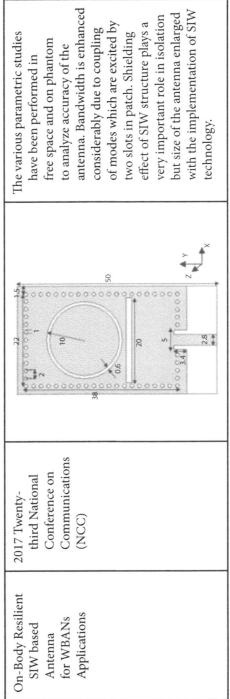 | The various parametric studies have been performed in free space and on phantom to analyze accuracy of the antenna. Bandwidth is enhanced considerably due to coupling of modes which are excited by two slots in patch. Shielding effect of SIW structure plays a very important role in isolation but size of the antenna enlarged with the implementation of SIW technology. |

4.3   Proposed Work

4.3.1   Geometry of Antenna

This antenna is fabricated on a substrate with dielectric constant ($\varepsilon r = 4.4$) and height 1.6 mm. Antenna structure along with its dimensions is depicted in Figure 4.7. On the top side of the substrate, the microstrip feeding line is etched. The width of the feeding line is 4 mm. The ground plane is printed on the bottom side of the substrate and the central part of it is cut by a rectangular shaped slot. A SIW is introduced to increase the gain and improve the compactness of the proposed antenna also the impedance bandwidth is improved. Based on the optimized parameters in Table 4.1 prototype is designed [14, 15, 18].

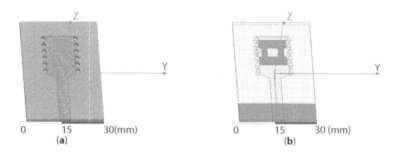

Figure 4.7 (a) Front side of proposed antenna. (b) Back side of proposed antenna.

Table 4.1 Design parameter of proposed antenna.

Antenna parameter	Dimension
Substrate length (x)	20.5 mm
Substrate width (y)	14 mm
Substrate FR4 (permittivity)	4.4
Radius of via (Rvia)	0.8 mm
Feed Width (feedy)	4 mm
Feed length (feedx)	17.5 mm
Ground slot length (gndx)	8 mm

4.3.2   Parametric Analysis

4.3.2.1   *Effect of Radius of Vias*

Variation of return loss with resonant frequency for the different value of via from 0.6 to 1.0 mm is exhibited in Figure 4.8. The radius of via effects the impedance matching as well as gain. With the increase in radius of via its gain deteriorates as shown in Figure 4.9.

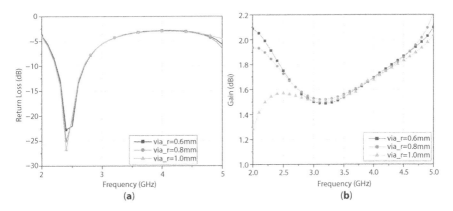

Figure 4.8 (a) and (b): Effect of variation of radius of via on return loss and gain of antenna.

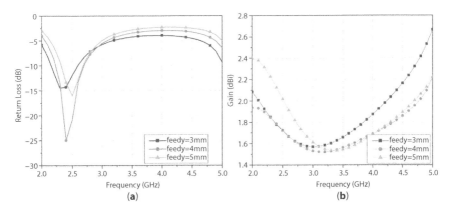

Figure 4.9 (a) and (b): Effect of variation of length of feed on return loss and gain of antenna.

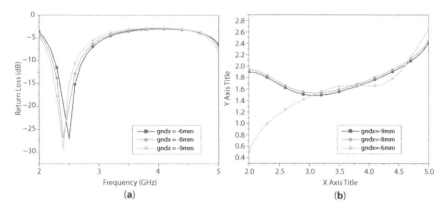

Figure 4.10 (a) and (b): Effect of defected ground structure on the return loss and gain of proposed antenna.

4.3.2.2   Effect of Length of Feed

With the increase or decrease in the width of feed mismatching occurs that highly affects the return loss and gain. Variation of feedy from 3 to 5 mm, the optimization results feedy = 4 mm gives the best impedance matching.

4.3.2.3   Effect of Length of Conductive Portion of Ground

Defected ground structure is one of the popular techniques that is used to enhance the patch antenna performance by increasing the gain and bandwidth, lowering cross polarization as shown in Figure 4.10. Variation in gndx shifts the resonance frequency, gndx = −6 mm shifts the resonance frequency towards right side also reduce the gain. gndx = −8 mm & −9 mm are depicted in the graph near to each other. After optimization gndx = −8 mm gives the required results with 95% efficiency [26, 27].

4.4   Result

The electromagnetic field remains confined inside the SIW, so there is no leakage of radiation. A rectangular patch is printed on the substrate with DGS along with proposed metallic via array. Figures 4.11 (a) and (b) depict the input impedance and gain of the proposed structure. Integration of metallic via array along with rectangular patch balances the capacitive reactance with inductive reactance to maintain the resonance frequency. The simulated gain attains the peak value around 1.9 dBi at 2.4 GHz.

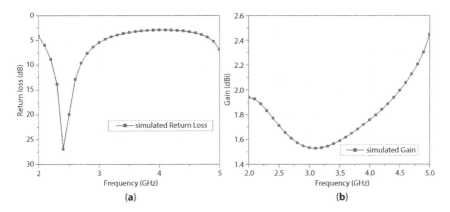

Figure 4.11 (a) and (b): Return loss and gain of proposed antenna.

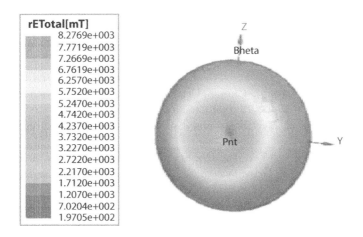

Figure 4.12 3D plot of the proposed antenna.

To describe the radiation characteristics of the proposed antenna, the simulated 3D radiation plots are shown in Figure 4.12.

4.5 Conclusion

A new design technique to implement SIW antenna is presented. A modified rectangular patch along with defected ground structure gives an omnidirectional pattern. The parametric study of the equivalent circuit model reveals the relationship between design dimensions and the coupling between cavity modes and feed line which serves as design guideline to achieve optimum performance of the antenna. The planar SIW based cavity backing of the slot antenna improves gain, FTBR at each resonance while maintaining low profile, compact design configuration which makes it suitable for modern wireless communication application.

References

1. Lai, A., Caloz, C., Itoh, T., Composite right/left-handed transmission line metamaterials. *IEEE Microw. Mag.*, 5, 3, 34–50, 2004.
2. Chhabra, B., Agrawal, T., Sharma, R., Dual Band SIW Bowtie Antenna for X band Applications. *IJCE*, 6, 1, Jan–June 2018, ISSN 2320-8996.
3. Means, D.L. and Kwok, W., *Evaluating Compliance with FCC Guidelines for Human Exposure to Radiofrequency Electromagnetic Fields, Supplement C (edition 01-01) to OET Bulletin 65 (Edition 97-01)*, Federal Communications Commission Office of Engineering & Technology, June 2001.
4. Chaturvedi, D. Dept. of ECE, NIT Trichy and Raghavan, S Dept. of ECE, NIT Trichy., On-Body Resilient SIW based Antenna for WBAN Applications. *2017 Twenty-Third National Conference on Communications (NCC)*.
5. Kang, D.-G., Kwon, J., Choi, J., Design of a Dual-Band Antenna for WBAN Applications. *The 2015 International Workshop on Antenna Technology*, Department of Electronics and Communications Engineering Hanyang University.
6. Kong, F., Qi, C., Lee, H., Durgin, G.D., Ghovanloo, M., Antennas for Intraoral Tongue Drive System at 2.4 GHz: Design, Characterization, and Comparison. *IEEE Trans. Microw. Theory Tech.*, 99, 1–10, January 2018.
7. Yalduz., H., Koç, B., Kuzu, L., Turkmen, M., An ultra-wide band low-SAR flexible metasurface-enabled antenna for WBAN applications. *Appl. Phys. A*, 125, 9, 609, 2019.
8. Ha, J., Kwon, K., Choi, J., Compact zeroth-order resonant antenna for implantable biomedical service applications. *Electron. Lett.*, 47, 23, 1267–1269, 2011.
9. Lee, J., Kwak, S.I., Lim, S., Wrist-wearable zeroth-order resonant antenna for wireless body area network applications. *Electron. Lett.*, 47, 7, 431–433, 2011.
10. Lacik, J. and Mikulasek, T., Circular Ring-Slot Antenna Fed by SIW for WBAN Applications. *2013 7th European Conference on Antennas and Propagation (EuCAP)*, Dept. of Radio electronics Brno University of Technology Brno, Czech Republic.
11. Bozzi, M., Georgiadis, A., Wu, K., Review of substrate-integrated waveguide circuits and antennas. *IET Microw. Antenna. P.*, 5, 8, 909–920, 2011.
12. NesaSudhaa, M., Benittaba, S.J., Design of a Dual-Band Antenna for WBAN Applications, Intell. Decis. Technol., 10, 203–209, 2016.

13. Lajevardi, M.E. and Kamyab, M., Ultra miniaturized Metamaterial-Inspired SIW Textile Antenna for Off-Body Applications. *IEEE Antennas Wirel. Propog. Lett.*, 16, 3155–3158, 2017.

14. Gao, G., Hu, B., Wang, S., & Yang, C. Wearable planar inverted-F antenna with stable characteristic and low specific absorption rate. *Microwave and Optical Technology Letters*, 60(4), pp. 876–882, 2018.

15. Hall, P.S. and Hao, Y., *Antennas and Propagation for Body-Centric Wireless Communications*, Artech House, Norwood, Mass, USA, 2006.

16. Moro, R., Agneessens, S., Rogier, H., Bozzi, M., Circularly-polarised cavity-backed wearable antenna in SIW technology. *IET Microw. Antenna. P.*, 12, 1, 127–131, 2017.

17. Delgado-Castillo, R.F.M. and Rodríguez-Solís, R.A., *Design of a V-Band Wideband Circularly Polarized Microstrip Patch Array Element for Applications in Body Area Networks*.

18. Youngtaek Hong, S., Tak, J., Choi, J., An All-Textile SIW Cavity-Backed Circular Ring-Slot Antenna for WBAN Applications. *IEEE Antennas Wirel. Propag. Lett.*, 15, 210–213, 2016.

19. Yan, S., Soh, P.J., Vandenbosch, G.A.E., Dual-Band Textile MIMO Antenna Based on Substrate-Integrated Waveguide (SIW) Technology. *IEEE Trans. Antennas Propag.*, 63, 11, 4640–4647, Nov 2015.

20. Yue, T. and Werner, D.H., A Compact Dual-Band Antenna Based on SIW Technology. *IEEE Trans. Antennas Propag.*, 779–780, 2018.

21. Kumar, A., Albreem, M.A., Gupta, M., Alsharif, M.H., Kim, S., Future 5G Network Based Smart Hospitals: Hybrid Detection Technique for Latency Improvement. *IEEE Access*, 8, 153240–153249, 2020.

22. Kumar, A., Gupta, M., Le, D.N., Aly, A.A., PTS-PAPR Reduction Technique for 5G Advanced Waveforms Using BFO Algorithm. *Intell. Autom. Soft Co.*, 27, 3, 713–722, 2021.

23. Meena, K., Gupta, M., Kumar, A., Analysis of UWB Indoor and Outdoor Channel Propagation. *2020 IEEE International Women in Engineering (WIE) Conference on Electrical and Computer Engineering (WIECON-ECE)*, IEEE, pp. 352–355, 2020.

24. Gupta, M., Chand, L., Pareek, M., Power preservation in OFDM using selected mapping (SLM). *J. Stat. Manage. Syst.*, 22, 4, 763–771, 2019.

25. Arun Kumar and Manisha Choudhary, "Dual band modified Split-Ring Resonator microstrip antenna for wireless applications, National Academy Science Letters, vol. 43, pp. 237-240, 2019. doi: 10.1007/s40009-019-00845-7, 2019

26. Hong, Y., Tak, J., Choi, J., An All-Textile SIW Cavity-Backed Circular Ring-Slot Antenna for WBAN Applications. *IEEE Antennas Wirel. Propag. Lett.,* 15, 1995–1999, 2016.

27. Hazarika, B., Basu, B., & Kumar, J. (2018). A multi-layered dual-band on-body conformal integrated antenna for WBAN communication. AEU-International Journal of Electronics and Communications, 95, pp. 226–235, 2018.

5

Analysis of RF-DC Rectifier Input Impedance for the Appropriate Design of Matching Network for Wireless RF Energy Harvesters

Kamini Singh[1]*, Sanjeev Yadav[2], J.K. Deegwal[3] and M.M. Sharma[4]

[1]NPIU-MHRD, Department of Electronics & Communication Engineering, Government Women Engineering College, Ajmer, India
[2]Department of Electronics & Communication Engineering, Government Women Engineering College, Ajmer, India
[3]Government Women Engineering College,Ajmer, India
[4]Department of Electronics & Communication Engineering, Malaviya National Institute of Technology, Jaipur, India

Abstract

The environmental radio frequency (RF) energy is used as an alternate solution of power for the easily accessible and non-accessible (hazardous) wireless sensor nodes. To convert available RF energy into direct current, rectifiers are needed. The matching network is required to match antenna impedance and rectifier input impedance. The design of matching circuit depends on the input impedance of rectifier. Rectifier circuit is non-linear in nature as diodes and MOSFETs are used for rectification. Therefore, input impedance is a function of various parameters like frequency, input power and output load. Consequently, matching network circuit design will change and maximum energy cannot be harvested from the incoming power as received power is very low. So in this chapter, a study of input impedance of RF-DC rectifier with non-linear Schottky diode model has been carried out on CADENCE virtuoso software at three different frequencies, *viz.* 930 MHz, 2.45 GHz and 3.5 GHz. This work demonstrates a change in input impedance of RF-DC rectifier from 5.991-j345.25 Ω to 5.987-j71.48 Ω as frequency

**Corresponding author*: kamini.singh3@gmail.com

Arun Kumar, Manoj Gupta, Mahmoud A. Albreem, Dac-Binh Ha and Mohit Kumar Sharma (eds.)
Wearable and Neuronic Antennas for Medical and Wireless Applications, (85–104) © 2022 Scrivener Publishing LLC

increases from 930 MHz to 3.5 GHz. With output load resistance, input impedance shows very large variations at lower frequency range and small variations at high frequency range.

Keywords: Wireless sensor networks (WSN), radio frequency (RF), energy harvesters, RF-DC rectifiers, matching networks, Schottky diodes

5.1 Introduction

Over the last few years, the network of wireless sensor nodes (WSN) has opened up a new area of Internet of Things (IoT), which enables the connection and transfer of data between various nodes which are placed at easily accessible and/or non-accessible (hazardous) areas. These nodes find applications in various fields such as medical [1], safety [2, 3], environmental [4], industrial [5, 6], etc. In non-accessible areas, the examination of the performance of the nodes is not an easy task. Conventionally, these nodes are operated with the help of batteries whose performance gets deteriorated with time. For instance, a dead battery of a seismic sensor may lead to very destructive consequences. Instead of batteries, using environmental energy may provide a better and long-lasting alternate for both accessible and non-accessible sensor nodes.

"Energy Harvesting" means to scavenge energy from the environmental sources. It provides a way to eliminate the need of battery for energy storage and prolong the system lifetime of a WSN. In easy terms, it is a system that supplies small but unbounded energy solutions for WSN [7]. A variety of sources are available in environment for energy harvesting including solar power, thermal power, energy from vibrations, wind energy, RF energy, etc. The amount of energy that can be harvested depends on the environmental condition and the application. The density of energy harvested from different sources is given in Table 5.1 [8]. Solar energy sources are capable of producing huge amount of power, but it is not useful for indoor applications and it suffers from the absence of scalability. Thermal energy harvesters store energy from thermal gradients [9]. The main focus of the present chapter is energy harvesting from RF sources.

5.1.1 Need and Advantages of Energy Harvesters

Working of Wireless Sensors Nodes relies on batteries, which provide power to the device. But use of batteries has several limitations such as, it has limited life-span or limited storage capacity, the size of the battery is large compared to WSN, batteries require periodic charging and

Table 5.1 Power density and efficiency of different sources.

Energy source	Power density	Efficiency	Challenges
Light	100 mW/cm^2 (Outdoor) 100 µW/cm^2 (Indoor)	10–25%	Where access to direct sunlight like indoor area, forestry terrains is not possible, solar energy harvesting may not be a correct choice.
Thermal	60 µW/cm^2 (Human) 10 mW/cm^2 (Industrial)	0.1–3%	The efficiency of thermal energy harvesters is very less for temperature gradient less than 40 °C.
Vibrations	4 µW/cm^2 (human) (Hz) 800 µW/cm^2 (machines) (KHz)	25–50%	Vibrational energy is dependent on driving frequency and acceleration.
Radio Frequency	0.1 µW/cm^2 (GSM 900 MHz) 0.001 µW/cm^2 (Wi-Fi 2.4 GHz)	50%	Ambient RF energy levels are very low.

the replacement of dead battery is very difficult in hazardous environment [10]. So it becomes a very costly and cumbersome task to replace batteries. This is one of the most considerable obstacles to the widespread commercial adoption of WSN. So researchers are focusing on alternative methods of energy for longer lifetime of WSNs without any replacement.

Harvesting energy from environment can eliminate this dependency on batteries; and it has many advantages such as, it provide boundless supply, it is green and clean and readily available when required, it eliminates the problem of recharging and replacement of batteries,and it is suitable for numerous deployments at unreachable locations.

5.1.2 RF Energy Sources

The huge deployment of RF communication makes RF energy available everywhere. RF power is widely broadcast from numerous reliable

Table 5.2 Different RF sources.

Sources	Maximum power	Power density
AM	50 kW Base station	40 W/m² at 10 km distance
Mobile	0.5 W	0.4 mW/m² at 10 m distance
GSM	100 W Base station	8 W/m² at 1,000 m distance
Wi-Fi	1 W Wi-Fi router	0.8 mW/m² at 10 m distance

electromagnetic sources like FM radio systems, TV transmission, cell tower transmission, AM transmission, mobile phones, Wi-Fi transmission, etc. Also, it is less dependent on traffic and weather conditions.

Other advantage of RF energy is that it is continuously received from the huge number of radio frequency (RF) transmitters. The number of Mobile phones, Wi-Fi systems and routers, base stations, base transmitters, Laptops, have reached above billions [11]. In urban environment, many Wi-Fi routers can be seen at a single place. Each Wi-Fi router may emit 50–100 mW in all directions although this much of emitted power is not fully consumed. Power density of various RF sources is shown in Table.5.2 [12].

The energy harvested from source is in unconditioned form. It cannot be used in this form to power a WSN. So, first it needs to be converted into usable form by power management unit and then stored in storage unit like supercapacitors, thin film batteries, etc. This complete block of energy harvester, power management circuit and storage unit is called energy harvesting system.

This chapter aims at those points which affect the efficiency of energy harvesters. Section 5.2 presents the basic information of blocks of energy harvester with antenna. The selection criterion of matching network has been detailed in Section 5.3. The comparison of pi & L type matching network efficiency has been carried out on LTspice IV software and presented in Section 5.3. In Section 5.4 analysis of input impedance of RF-DC rectifier has been done on CADENCE virtuoso. At last, in Section 5.5 the conclusions are presented.

5.2 RF Energy Harvesting Processing Block

An RF energy harvester comprises of the following blocks: antenna, matching network, and rectifier (RF-DC) as shown in Figure 5.1. According to the application, DC–DC charge pump or Boost converters can be used.

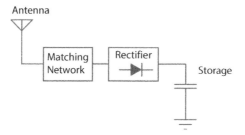

Figure 5.1 RF energy harvesting system.

Antenna is used for receiving the radio frequency signal. In the electrical model representation of antenna a source of AC voltage (V_a) is in series with impedance (Z_a) as shown in Figure 5.2. The impedance of antenna Z_a is expressed by Eq. (5.1), where R_a, real impedance is obtained by two resistances-R_l (material used for antenna) and R_s (electromagnetic wave radiation). The imaginary element X_a, on the other hand, depends on the antenna structure. Commonly used Z_a values are 50, 75, and 300 Ω for wireless systems, open dipole antenna and closed dipole antenna [13].

$$Z_a = R_a + X_a \qquad (5.1)$$

The AC voltage source amplitude be determined by the received power (P_r) and the impedance real component (R_a) and can be written as Eq. (5.2) [14].

$$V_a = 2\sqrt{2R_aP_r} \qquad (5.2)$$

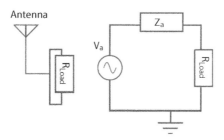

Figure 5.2 Antenna electrical model representation.

The received power by antenna is represented by Friss equation [15] as shown in Eq. (5.3).

$$P_r = \frac{P_t G_t G_r \lambda^2}{(4\pi r)^2} \tag{5.3}$$

where, P_r, P_t, G_t, G_r, λ, and r are received power, transmitted power, transmitting antenna gain, receiving antenna gain, wavelength of transmitting wave, the distance between transmitting and receiving antenna respectively. The losses or attenuations occur due to the travelling path of the wave are called as path loss. The path loss (P_l) of free space is given by Eq. (5.4) [16].

$$P_l = \frac{(4\pi r f)^2}{G_t G_r c^2} \tag{5.4}$$

The path loss depends on the distance between transmitter and receiver, transmission frequency (f), and gain of transmitter and receiver antenna. As frequency increases, received power is decreased. Hence it is easy to harvest energy at low frequency bands than high frequency bands. According to the received power, RF energy harvesters work can be categorized into low power and high power regions. In low power region, power level is from −30 to 0 dBm and high power region shows power levels from 0 to 30 dBm [16]. Energy can be easily stored in high power region but the difficulty arises when the input power is very low (in μW range). Harvested output power can be defined as Eq. (5.5) from Figure 5.1.

$$P_o = P_i \, \eta_{ant} \, \eta_{mn} \, \eta_{rect} \tag{5.5}$$

where P_o is output power of the energy harvester, P_i is input power received by antenna, η_{ant} is antenna efficiency, η_{mn} is matching network efficiency, and η_{rect} is rectifier efficiency. Consequently, all the blocks of RF energy harvester should work efficiently.

5.3   Matching Network & RF-DC Rectifier

Between the antenna and the load, matching network assists maximum power transfer. A matching network matches the impedance of source and load. The load impedance is defined by the rectifier and remaining circuit in the RF energy harvester. When there is a mismatch, reflection of the incident wave occurs andconsequently, efficiency is reduced. A matching network confirms transfer of maximum received power between the source and load by matching [17].

Matching network can be designed by using a single resistor, but it is not an appropriate solution since maximum power will be lost in the resistor. And in this matching only resistive part will be matched. Other matching networks are available which are based on the series and parallel combination of L (inductor) and C (capacitor), like pi, L and T type matching networks. These are the simplest form of matching networks. L-section includes series inductor with shunt capacitor or series capacitor with shunt inductor as shown in Figure 5.3.

The pi and T section matching network show less efficiency than L-section but provide larger bandwidth [18–20]. For the same input and output impedances, L and pi matching network design has been compared on LTSpice IV & Cadence software at 930 MHz and 2.45 GHz frequency. The results, shown in Figure 5.4, also represent that L type matching network provides larger output for the same input at higher frequencies. The reflection coefficient (S11) for L-matching & pi matching circuits are −48.8 dB and −34.72 dB at 2.45 GHz frequency and −50.398 dB and −34.60 dB at 930 MHz frequency as shown in Figure 5.5. It can be easily understood from the results that L-matching network provides highest frequency stability. Matching network design depends on the load resistance value

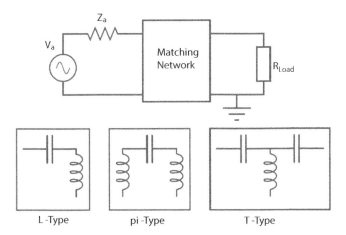

Figure 5.3 L, pi, T type matching circuits.

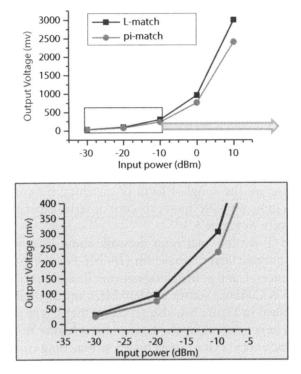

Figure 5.4 L & pi type matching circuits output voltage.

which is going to be matched for the maximum power transmission. In energy harvester, load is RF-DC rectifier and output load.

Diode is a basic component of rectification and it gives rectified output when applied or received input voltage is greater than knee or threshold voltage of the diode. Generally, diodes and CMOS based rectifiers are used in RF to DC rectifiers [16–18]. Due to low forward voltage, Schottky diodes (Commercially available diodes HSMS285×, HSMS 286×, SMS2761) are commonly used [21–25]. Furthermore, complementary metal-oxide-semiconductor (CMOS) has replaced the diodes in many reported works [25–28]. Fast switching and high integration level are the major advantages of MOSFET, but it requires a high value of threshold voltage and its operation at high frequency restricts the efficiency of energy harvesters. As frequency increases, efficiency decreases due to power loss occurring in the MOSFET from the reverse leakage current [29, 30]. So for low RF input power, Schottky diode based rectifiers deliver better efficiency than MOSFET.

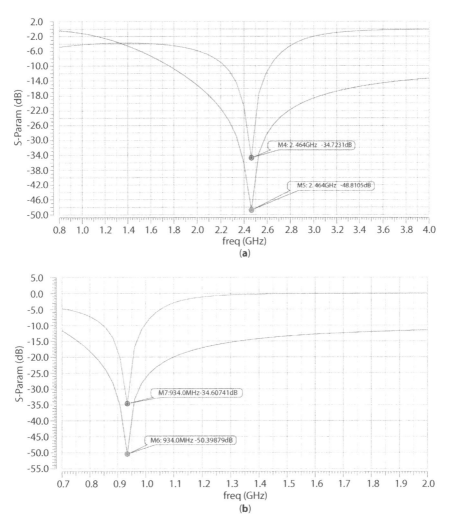

Figure 5.5 (a) L & pi type Matching Circuits S11 at 2.45 GHz (b) L & pi type Matching Circuits S11 at 930 MHz.

5.4   Study of Input Impedance of Rectifier

The selection of matching network circuit depends on the requirements such as high output or wider bandwidth, as mentioned in Section 5.3. Another important concern in the matching network design is output load impedance value. A constant output load impedance doesn't affect the design of matching network, but the efficiency of matching network

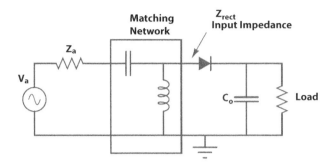

Figure 5.6 Rectifier input impedance.

changes with change in output impedance. In energy harvesters, load is rectifier input impedance as shown in Figure 5.6. In this work, a simple structure of Schottky diode based Dickson voltage multiplier has been used as a rectifier for RF energy harvester. The choice of rectifier depends on the input RF signal frequency and power.

A packaged Schottky diode is modeled as a nonlinear R_j (junction resistance) shunted by a non-linear C_j (junction capacitance) as shown in Figure 5.7. V_j is the voltage across the junction resistance. R_S is the current spreading resistance, L_p & C_p are packaging inductance and capacitance [26]. Capacitance and inductance differentiate the working of Schottky diodes at low and high frequencies.

The junction resistance R_j is given by Eq. (5.6) [26]

$$Rj = \frac{\eta\alpha}{I_s + I} \tag{5.6}$$

Where η, α, and I_s are the diode ideality factor, thermal voltage and diode reverse saturation current. Amperes current (I) changes with the input RF

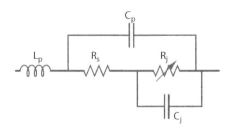

Figure 5.7 Schottky diode non-linear model [26].

power level P_{in} and it affects the junction resistance (R_j) and consequently, the overall impedance of the Schottky diode. From Figure 5.7 the overall impedance (Z_{rect}) of Schottky diode can be written as Eq. (5.7).

$$Z_{rect} = jwL_P + \frac{\left(\left\{\dfrac{Rj}{1+jwRjCj}+R_S\right\}\dfrac{1}{jwC_P}\right)}{\left(\left\{\dfrac{Rj}{1+jwRjCj}+R_S\right\}+\dfrac{1}{jwC_P}\right)} \quad (5.7)$$

For the exact evaluation of input impedance, one stage diode Dickson voltage multiplier RF-DC rectifier circuit has been designed in CADENCE, virtuoso as shown in Figure 5.8. In this circuit, non-linear model (shown in Figure 5.7) of HSMS2850 Schottky diode has been used at the place of D1 and D2 as it works better at low input power levels and at high frequency [27–30].

Spice parameters of Schottky diode HSMS-2850 are given in Table 5.3.

The simulation has been performed at three different frequencies of 930 MHz, 2.45 GHz and 3.5 GHz. The result of Figure 5.9 shows that impedance is varying from 5.991-j345.25 Ω to 5.987-j71.48 Ω as frequency increases from 930 MHz to 3.5 GHz.

The variation in input impedance of single Schottky diode (HSMS 2850) with load resistance can be seen from Figure 5.10. The results show that

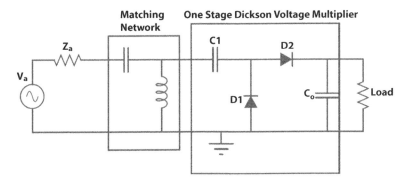

Figure 5.8 Complete Energy harvester with one stage Dickson multiplier.

Table 5.3 SPICE parameter of Schottky Diodes [27, 28, 30].

Parameter	Bv	C_{j0}	Eg	I_{BV}	I_S	N	R_s	PB (V_j)	PT(Xti)	M
HSMS-2850	3.8 V	0.18p	0.69 eV	300 μA	3 μA	1.06	25 Ω	0.35 V	2	0.5

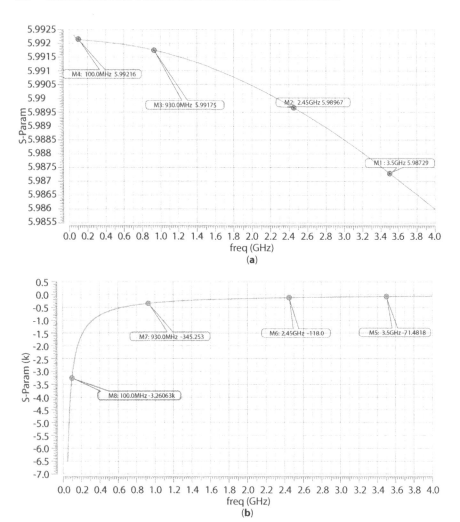

Figure 5.9 (a) real impedance variation at different frequencies, (b) imaginary impedance variation at different frequencies.

at low frequency, changes in impedance are significant and cannot be ignored, but at high frequencies it shows very small change. At 50 MHz frequency, real input impedance changes from 22.10 Ω to 12.084 Ω, and at 2.45 GHz frequency real input impedance change from 11.981 to 11.976 Ω, when load changes from 10 KΩ to 1 MΩ. This change in input impedance occurs due to the non-linear Schottky diode model, in which reactance of junction capacitances and inductances varies according to the frequencies (as shown in Eq. (5.7)).

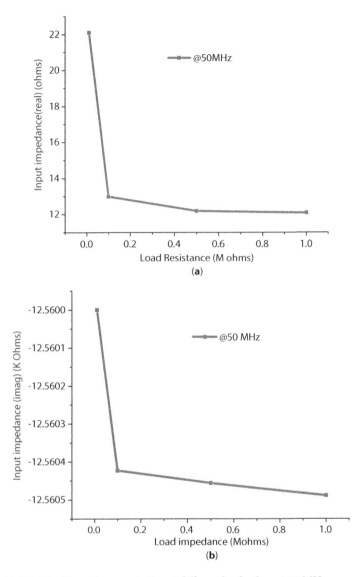

Figure 5.10 (a) Real impedance variation at different load values at 50 MHz; (b) Imaginary impedance variation at different load values at 50 MHz;   (*Continued*)

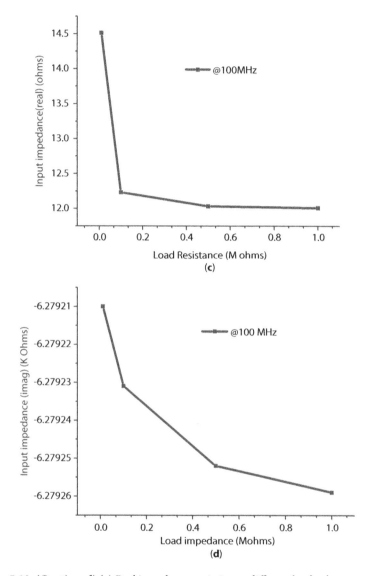

Figure 5.10 (**Continued**) (c) Real impedance variation at different load values at 100 MHz; (d) Imaginary impedance variation at different load values at 100 MHz;   (*Continued*)

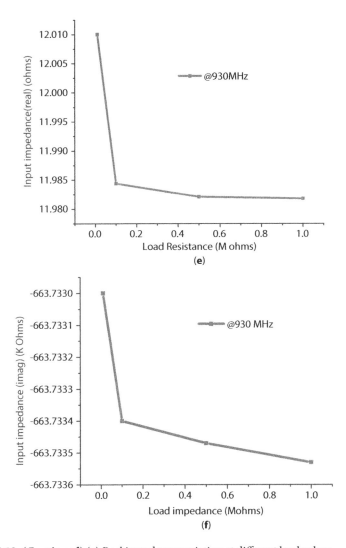

Figure 5.10 (**Continued**) (e) Real impedance variation at different load values at 930 MHz; (f) Imaginary impedance variation at different load values at 930 MHz;   (*Continued*)

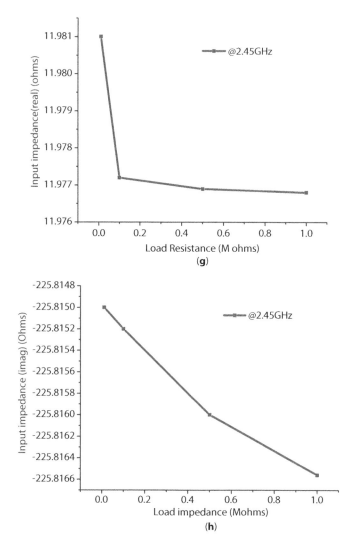

Figure 5.10 (**Continued**) (g) Real impedance variation at 2.45 GHz for different loads; (h) Imaginary impedance variation at 2.45 GHz for different loads.

5.5   Conclusion

In this chapter, various processing blocks such as antenna, matching network, RF-DC rectifier of RF energy harvester has been explained. A comparison of L and pi matching network has been done. The results show that for high efficiency, L-matching and for wide bandwidth, pi matching network give best responses. For the design of matching network, the effects of variation in the input impedance of RF-DC rectifier due to various components have been presented. From the analysis, it can be inferred that at higher frequencies rectifier shows very low value of input impedance. This result signifies low sensitivity of the rectifier. The input impedance values of single Schottky diode at 930 MHz and 2.45 GHz frequencies at 1 MΩ load are 11.9818-j 663.733 and 11.9768-j 225.815 respectively. The impedance values changes to 5.991-j345.25 and 5.9896-j118, when one stage of Dickson voltage RF-DC rectifier is used. With frequency, reactance impedance changes sharplyand very small change occurs in real impedance, and with variations of load impedance, a significant change in input impedance of RF-DC rectifier occurs at low frequencies. All the aforementioned factors should be considered in the design of matching network so that maximum power transfer from antenna to theRF-DC rectifier takes place.

Acknowledgment

The authors thankfully acknowledge the funding support received from the National Project Implementation Unit (NPIU), MHRD, Government of India. Above mentioned work is a part ofproject that was sanctioned under the TEQIP Collaborative research scheme, CRS ID: 1-5727948711.

References

1. Stankovic, J.A., Cao, Q., Doan, T., Fang, L., He, Z., Kiran, R., Lin, S., Son, S., Stoleru, R., Wood, A., Wireless Sensor Networks for In-Home Healthcare. Potential and Challenges. *High Confidence Medical Devices Software and systems, HCMDSS*, 2005.
2. Kumar, K.A., IMCC Protocol in Heterogeneous Wireless Sensor Network for High Quality Data Transmission in Military Applications. *2010 First International Conference on Parallel, Distributed and Grid Computing (PDGC 2010)*, Solan, pp. 339–343, 2010, doi: 10.1109/PDGC.2010.5679973.

3. Torfs, T., Sterken, T., Brebels, S., Santana, J., van den Hoven, R., Spiering, V., Low Power Wireless Sensor Network for Building Monitoring. *IEEE Sens. J.*, 13, 3, 909–915, March 2013. doi: 10.1109/JSEN.2012.2218680.

4. Martinez, K., Padhy, P., Riddoch, A., Ong, H.L., Hart, J.K., Glacial Environment Monitoring using Sensor Networks. *REALWSN'05*, June 21–22, 2005, doi: 1-58113-000-0/00/0004.

5. Akhondi, M.R., Talevski, A., Carlsen, S., Petersen, S., Applications of Wireless Sensor Networks in the Oil, Gas and Resources Industries. *24th IEEE International Conference on Advanced Information Networking and Applications (AINA '10)*, pp. 941–948, April 2010, doi: 10.1109/AINA. 2010.18.

6. Płaczek, B., Selective Data Collection in Vehicular Networks for Traffic Control Applications. *Transp. Res. Part C: Emerg. Technol.*, 23, 14–28, 2012, doi: DOI: 10.1016/j.trc.2011.12.007.

7. Basagni, S., Naderi, M.Y., Petrioli, C., Spenza, D., Wireless Sensor Networks with Energy Harvesting, in: *Mobile Ad Hoc Hetworking*, vol. 1, pp. 701–736, https://doi.org/10.1002/9781118511305.ch20.

8. DeMil, P., Jooris, B., Tytgat, L., Catteeuw, R., Moerman, I., Demeester, P., Kamerman, A., Design and Implementation of a Generic Energy-Harvesting Framework Applied to the Evaluation of a Large-Scale Electronic Shelf-Labeling Wireless Sensor Network. Hindawi Publishing Corporation. *EURASIP J. Wirel. Commun. Netw.*, Article ID 343690, 1–12, 2010. doi:10.1155/2010/343690.

9. Gilbert, J.M. and Balouchi, F., Comparison of Energy Harvesting Systems for Wireless Sensor Networks. *Int. J. Autom. Comput.*, 05, 4, 334–347, October 2008. doi: 10.1007/s11633-008-0334-2.

10. Roundy, S., Steingart, D., Frechette, L., Wright, P., Rabaey, J., Power Sources for Wireless Sensor Networks. *Wireless Sensor Networks. EWSN 2004. Lecture Notes in Computer Science*, vol. 2920, Springer, Berlin, Heidelberg, 2004, https://doi.org/10.1007/978-3-540-24606-0_1.

11. Sidhu, R., Ubhi, J., Aggarwal, A., A Survey Study of Different RF Energy Sources for RF Energy Harvesting. *International Conference on Automation, Computational and Technology Management (ICACTM)*, 2019, doi: 978-1-5386-8010-0/19.

12. Pavone, Buonanno, A., Urso, M.D., Corte, F.D., Design Considerations for Radio Frequency Energy Harvesting Devices. *Prog. Electromagn. Res. B*, 45, 19–35, 2012. doi: 10.2528/PIERB12062901.

13. Xie, F., Yang, G.M., Geyi, W., Optimal Design of an Antenna Array for Energy Harvesting. *IEEE Antennas Wirel. Propag. Lett.*, 12, 155–158, 2013. doi: 10.1109/LAWP.2013.2243697.

14. Gasulla, M., Jordana, J., Robert, F., Berenguer, J., Analysis of the Optimum Gain of a High-Pass L-Matching Network for Rectennas. *Sensors*, 17, 1712, 2017. doi: 10.3390/s17081712.

15. Pozar, D.M., *Microwave Engineering*, 3rd edition, Wiley, 1998. http://mwl. diet.uniroma1.it/people/pisa/RFELSYS/MATERIALE%20INTEGRATIVO/ BOOKS/Pozar_Microwave%20Engineering (2012).pdf

16. Divakaran, S.K., Krishna, D.D., Nasimuddin, R.F., Energy Harvesting Systems: An Overview and Design Issues. *Int. J. RF Microw. C. E.*, 29, 1, p.e21633, 2018. doi/10.1002/mmce.21633.

17. Han, Y. and Perreault, D.J., Analysis and Design of High Efficiency Matching Networks. *IEEE Trans. Power Electron.*, 21, 5, 1484–1491 September 2006, doi: 10.1109/TPEL.2006.882083.

18. Agrawal, S., Kumar Pandey, S., Singh, J., Parihar, M.S., Realization of Efficient RF Energy Harvesting Circuits Employing Different Matching Technique. *15th International Symposium on Quality Electronic Design*, 2014, doi: 10.1109/ISQED.2014.6783403.

19. Singh, G., Ponnaganti, R., Prabhakar, T.V., Vinoy, K.J., A Tuned Rectifier for RF Energy Harvesting from Ambient Radiations. *Int. J. Electron. Commun. (AEÜ)*, 67, 7, 564–569, 2013. http://dx.doi.org/10.1016/j.aeue.2012.12.004.

20. Alneyadi, F., Alkaabi, M., Alketbi, S., Hajraf, S., Ramzan, R., 2.4GHz WLAN RF Energy Harvester for Passive Indoor Sensor Nodes. *IEEE-ICSE2014 Proc*, Kuala Lumpur, Malaysia, 2014, doi:10.1109/SMELEC.2014.6920900.

21. AbdKadir, E. and Hu, P.A., A Power Processing Circuit for Indoor Wi-Fi Energy Harvesting for Ultra-Low Power Wireless Sensors. *Appl. Sci.*, 7, 8, 827, August 2017. doi: 10.3390/app7080827.

22. Karolak, D., Taris, T., Deval, Y., Béguéret, J.B., Mariano, A., *Design Comparison of Low-Power Rectifiers Dedicated to RF Energy Harvesting*, In *2012 19th IEEE International Conference on Electronics, Circuits, and Systems (ICECS 2012)* (pp. 524-527). IEEE2012, doi: 978-1-4673-1260-8/12.

23. Ouda, M.H., Arsalan, M., Marnat, L., Shamim, A., Salama, K.N., 5.2-GHz RF Power Harvester in 0.18- m CMOS for Implantable Intraocular Pressure Monitoring. *IEEE Trans. Microw. Theory Tech.*, 61, 5, 2177–2184 May 2013. doi: 10.1109/TMTT.2013.2255621.

24. Rosli, M.A., Murad, S.A.Z., Norizan, M.N., Ramli, M.M., Design of RF to DC Conversion Circuit for Energy Harvesting in CMOS 0.13-μm Technology. *4th Electronic and Green Materials International Conference 2018 (EGM 2018)*, doi: 10.1063/1.5080902.

25. Kang, S.-M. and Leblebici, Y., *CMOS Digital Integrated Circuits*, McGraw-Hill, 1996.

26. Nimo, A., Beckedahl, T., Ostertag, T., Reindl, L., Analysis of Passive RF-DC Power Rectification and Harvesting Wireless RF Energy for Micro-watt Sensors. *AIMS Energy*, 3, 2, 184–200, 2014. doi: 10.3934/energy.2015.2.184.

27. *Agilent HSMS-285x Series Surface Mount Zero Bias Schottky Detector Diodes.* Agilent Technologies, June 22, 2005, https://datasheet.octopart.com/HSMS-2852-BLKG-Avago-datasheet-26413.pdf

28. *Surface Mount Microwave Schottky Detector Diodes Technical Data HSMS-2850 Series, HSMS-2860 Series,* AvagoTechnologies, Hewlett Packard, 1998,

https://www.datasheet.live/index.php?title=Special:PdfViewer&url=https%3A%2F%2Fpdf.datasheet.live%2F73f570c%2Favagotech.com%2F-HSMS-2850-TR1.pdf.

29. Singh, Deegwal, J.K., Sharma, M.M., Yadav, S., Performance Comparison of Schottky Diode models for RF Energy Harvesting. *Int. J. Eng. Res. Appl.*, ISSN: 2248-9622, Vol. 10,6, (Series-VI), 51–57, June 2020. doi: 10.9790/9622-100606515.

30. Mouapi, A., Hakem, N., Kandil, N., *A Performance Analysis of Schottky Diode to support RF Energy Harvesting*, IEEE, 2019, doi: 978-1-7281-0692-2/19.

6

Secured Schemes for RF Energy Harvesting Mobile Computing Networks with Multiple Antennas Based on NOMA and Access Points Selection

Van-Truong Truong[1,2*], Anand Nayyar[3,4] and Dac-Binh Ha[1,2]

[1]Faculty of Electrical-Electronic Engineering, School of Engineering and Technologies, Duy Tan University, Da Nang, Vietnam
[2]Institute of Research and Development, Duy Tan University, Da Nang, Vietnam
[3]Graduate School, Duy Tan University, Da Nang, Vietnam
[4]Faculty of Information Technology, Duy Tan University, Da Nang, Vietnam

Abstract

The Internet of Things (IoT) has explosive growth in many fields, but IoT devices often have minimal computation resources, which reduces the QoS experience. Mobile Edge Computing (MEC) is a promising technique to address this issue when it responds well to mobile applications requiring computation-intensive with latency-critical. In this chapter, we investigate RF energy harvesting MEC networks, in which a power station provides RF energy to a multi-antenna user for tasks offloading to the access point (AP) based on NOMA over Nakagami-m fading. User deploys selection combining (SC) or maximum ratio combining (MRC) to harvest RF energy and transmit antenna selection (TAS) or maximum ratio transmission (MRT) for offloading. We investigate four schemes, namely SC-TAS A, SC-TAS B, SC-MRT, and MRC-MRT, based on AP selection. Accordingly, we obtain closed-form expressions in terms of successful computation probability (SCP). Finally, a novel framework is proposed to maximize the SCP using two metaheuristic algorithms. Monte-Carlo simulations are used to verify the analysis.

Keywords: Mobile edge computing, RF energy harvesting, SC/MRC, TAS/MRT, NOMA, antenna, GA, PSO

Corresponding author: truongvantruong@dtu.edu.vn

Arun Kumar, Manoj Gupta, Mahmoud A. Albreem, Dac-Binh Ha and Mohit Kumar Sharma (eds.)
Wearable and Neuronic Antennas for Medical and Wireless Applications, (105–136) © 2022 Scrivener Publishing LLC

6.1   Introduction

In recent years, the academic and industrial communities have made great efforts to realize and commercialize 5G. The 5G networks not only supports communication functions, but also offers three other functions, including compute, control, and content distribution [1]. Accordingly, a lot of new applications with requirements such as very low latency, high bandwidth, high reliability, and dense user count, i.e., virtual/augmented reality (VR/AR), autonomous vehicle, Tactile Internet, and Internet of Things (IoT) were born [2]. However, many of the wireless devices connected to the above systems do not meet the computing, energy and storage requirements. It led to the advent of mobile edge computing (MEC). In the MEC paradigm, telecommunication and IT cloud services are combined to provide cloud computing abilities in radio access networks in mobile users' proximity [3–5]. Thus, the wireless devices with the constraints on energy and computing resources can offload tasks to the MEC server deployed at the base station (BS), wireless access points (AP), or even nearby laptops.

Currently, the problem of allocating tasks to users in the MEC network has attracted significant research, when mobile users often have minimal energy resources, and this energy source is not enough for them to have executing computation-intensive and latency-critical applications.

To solve this problem, radio frequency energy harvesting (RF EH) is considered a potential technique when integrated into the MEC network. This technique improves the user's energy efficiency and increases their battery life-time as well as ensures network coverage [6–8]. Several prior studies have shown that using this technique can improve system performance [9–12]. In work [9], Chen *et al.* proposed the multi-user multi-task computation offloading problem of the wireless EH-enabled devices. The authors use the Lyapunov optimization method to provide optimal energy harvest policy and task offload schedule. Zhang *et al.* in [10] use the theoretic game to evaluate the effect of the RF EH technique on the task offload model in MEC networks. Another example is Wu *et al.* in work [12], which solved the problem of resource allocation in EH-enabled MEC server and considered load balancing condition.

Furthermore, diversity techniques applied at the receiver, such as selection combining (SC) and maximal ratio combining (MRC) can help improve energy harvesting process efficiency [13, 14]. Alcaraz *et al.* in work [14] investigated the single-input multiple-output (SIMO) model wireless systems over Rayleigh fading and focused on the maximization the energy efficiency. Research has shown that MRC is more efficient in energy collection than SC, but get a higher complexity.

In addition, the non-orthogonal multiple access (NOMA) technique can be integrated into the MEC network to improve spectral efficiency and throughput compared to traditional OMA [15, 16] techniques. NOMA can support a large number of users with very high speed in the same frequency and time resource. Thus, it is evident that the combination of NOMA and RF EH techniques in MEC network promises to solve many problems such as energy consumption minimization [17–19], execution latency minimization [20–22] and revenue maximization [17, 18, 20, 23].

Zeng *et al.* in [17] proposed a multi-user MEC network model, in which each user can harvest energy from the wireless AP and offload their task to the MEC server located there. The authors proposed an algorithm aimed to find the optimal power transfer time allocation for the binary offloading problem. The system performance is evaluated through the parameters of the sum rate, the number of offloading users and is superior to when applying TDMA technology. Zhou *et al.* in work [18] proposed the NOMA-MEC model applied downlink wireless power transfer to multiple users. The system uses a non-linear EH model and can operate under two schemes, namely binary offloading and partial offloading. The authors proposed two algorithms, the iterative algorithm and an alternative optimization algorithm based on SCA, to optimize system performance. The work [20] by Rauniger *et al.* proposed RF network EH NOMA-MEC cooperation. In this work, one user transmits its signal to the base station; however, due to too much interference in the transmission channel, this process cannot be executed directly. Therefore, the user needs the help of the IoT node to act as the DF relay to deliver the signal. The authors used the Golden section search algorithm to determine the optimal power allocation coefficient for the system.

In addition, the application of multi-antenna and diversity techniques such as transmit antenna selection (TAS) and maximal ratio transmission (MRT) can also increase outage probability and secrecy performance [24–26].

To the best of our knowledge, there is no prior work that considers the NOMA MEC system that integrated RF EH technique. In this study, we investigate the RF EH NOMA MEC network, in which one multi-antenna user will use SC/MRC technique to get energy from a power station, then the user uses the TAS/MRT technique to offload its task to the MEC servers located at the wireless AP.

The main contributions of our work are as follow:

  – We investigate an RF EH NOMA MEC network, in which one multi-antenna user harvests energy from the power station using downlink NOMA and communicate with two wireless APs using uplink NOMA over Nakagami-*m* fading. We propose protocols for the system by combining SC/MRC for energy harvesting and TAS/MRT for task offloading. From there, we derive the closed-form expressions of the SCP corresponding these schemes using Gaussian–Chebyshev quadrature method.

– We propose two algorithms based on genetic algorithm (GA) and Particle Swarm Optimization (PSO) to determine the optimal set of parameters (including power allocation coefficient, data allocation coefficient and time switching ratio) to optimize the SCP for the system.

– Finally, we evaluate the performance of the in term of SCP under critical parameters, including task length, bandwidth, power allocation coefficient, data allocation coefficient, and time switching ratio, and operating frequency for the more understanding of the system's operation.

The rest of this chapter is organized as follows: Section 6.2 described the system model. Section 6.3 described the detailed derivations of the closed-form expression of the successful computation probability used to investigate the considered system performance. Section 6.4 provided the numerical results and some discussions. Finally, Section 6.5 concludes the paper.

6.2   System and Channel Models

We define in Table 6.1 the notations will be used in the next sections of the chapter.

In this chapter, we investigate the RF EH NOMA MEC network over Nakagami-m fading channel, as shown in Figure 6.1. Specifically, a multi-antenna user (U) harvests RF energy from the power station (P) and offload its task to two MEC access points (APs). Assuming all of the devices in the network operate in half-duplex mode, P and the APs are single-antenna devices, AP_A is trusted AP while AP_B is un-trusted AP. Assume that U needs to process a task of L bit length; however, due to energy limitations, U cannot compute locally. Therefore, U uses either SC or MRC techniques to harvest energy from P, and uses whole energy to offload the task to APs using TAS or MRT techniques. We assume that the task-input bits are bit-wise independent and can be arbitrarily divided into different groups [27–30]. Due to the security constraint of the system, AP_A will be prioritized for offloading, and AP_B is used to ensure system performance if necessary [27]. Thus, the L bit can be divided into Task 1 with L_1

Table 6.1 Notations.

Notation	Meaning
h_{0n}	Channel coefficient from P to antenna n of the user U
h_{n1}	Channel coefficient from antenna n of the U to AP_A
h_{n2}	Channel coefficient from antenna n of the U to AP_B
m_{PU}, m_{UA}, m_{UB}	Nakagami-m shape factor of the link from P to U, U to AP_A, and U to AP_B
N	Number of antennas at the user U
α	Time switching ratio
b	Power allocation coefficient
η	Energy conversion efficiency
L	Length of the task
T	Maximum allowed time delay
f_1, f_2	AP_A and AP_B operating frequency
c_i	Number of CPU cycles needed to compute one input bit
Pr_s	Successful computation probability

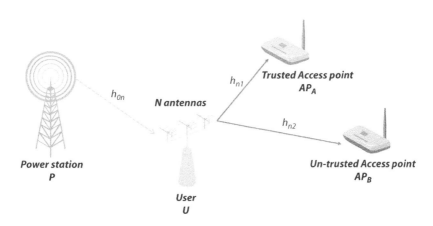

Figure 6.1 System model for RF EH NOMA MEC network.

bit requiring high confidential and offloaded to AP_A, while the remaining task $L_2 = L - L_1$ bits, namely Task 2, will be offloaded to AP_B. The time flow-chart for this considered system is given as Figure 6.2.

Let h_{0n}, h_{n1}, and h_{n2} $(1 \leq n \leq N)$ denote the channel coefficients of the links from power station P to the antenna n at user U, the antenna n at user U to AP_A, and the antenna n at user U to AP_B, respectively. Herein, h_p $(P \in \{0n, n1, n2\})$ is an independent and identically distributed (i.i.d) random variable following Nakagami-m distribution with parameter m_Q $(Q \in \{PU, UA, UB\})$ and the mean value $= E\{|h_p|^2\}$, where $E\{.\}$ stands for the expectation operator.

Before performing the communication process, U chooses one of the four schemes described in Table 6.2 for the system to work. In the case of using the optimal method, U uses one of two optimization algorithms, Algorithm 2 or Algorithm 3, to determine the system's optimal set of parameters.

We proposed the APS protocol for the RF EH NOMA MEC network consisting of four phases as:

- Phase 1: We call this phase the EH phase, where U will use one of the two diversity techniques SC or MRC to harvest energy from P in the period $\tau_0 = \alpha T$, where α denotes the time switching ratio and T denotes the maximum allowed

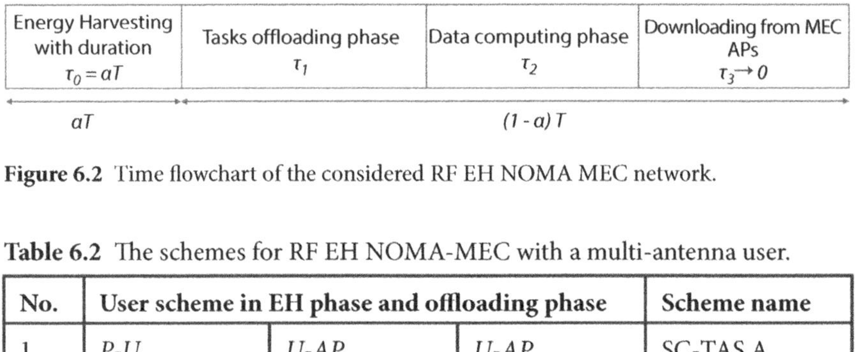

Energy Harvesting with duration $\tau_0 = \alpha T$	Tasks offloading phase τ_1	Data computing phase τ_2	Downloading from MEC APs $\tau_3 \rightarrow 0$
αT		$(1-\alpha)T$	

Figure 6.2 Time flowchart of the considered RF EH NOMA MEC network.

Table 6.2 The schemes for RF EH NOMA-MEC with a multi-antenna user.

No.	User scheme in EH phase and offloading phase			Scheme name
1	P-U SC	U-AP$_A$ TAS A	U-AP$_B$ TAS A	SC-TAS A
2	P-U SC	U-AP$_A$ TAS B	U-AP$_B$ TAS B	SC-TAS B
3	P-U SC	U-AP$_A$ MRT	U-AP$_B$ MRT	SC-MRT
4	P-U MRC	U-AP$_A$ MRT	U-AP$_B$ MRT	MRC-MRT

time delay. If using the optimal method, U uses the optimal time switching ratio α^*, which has been determined thanks to the optimal algorithm.

- Phase 2: We call this phase Offloading phase. According to the channel state information (CSI) and the diversity TAS or MRT technique is selected, U will choose the AP to offloadits task in duration τ_1 using NOMA as Algorithm 1with optimal power allocation coefficientsand data allocation coefficients.
- Phase 3: We call this phase Computing phase. After the communication is successful, AP_A and AP_B will use the SIC technique to decode the received signal, then compute the corresponding tasks in duration τ_2.
- Phase 4: We call this phase the Downloading phase. After the calculation is successful, the APs will feedback the calculation result to U in duration τ_3. These results are usually straightforward and concise and can be transmitted in a short time; therefore, assume τ_3 much smaller than τ_0, τ_1 and τ_2 and can be neglected [31, 32].

Algorithm 1: Access Point Selection (APS) Algorithm

Input:

      Scheme name, CSI, L, L_1, L_2

Output: none

Procedure APS()

---

1. **if** (TAS scheme) **then** $Y = \max\limits_{1 \leq n \leq N}\left\{|h_{n1}|^2\right\}, Z = \max\limits_{1 \leq n \leq N}\left\{|h_{n2}|^2\right\}$

2. **else if** (MRT scheme) **then** $Y = \sum\limits_{n=1}^{N}|h_{n1}|^2, Z = \sum\limits_{n=1}^{N}|h_{n2}|^2$

3. **end if**

4. **if** $(Y > Z)$ **then**

5. AP_A selected

6. U offload L-bit task to AP_A

7. go to **exit**

8. **else**

9. U divides L-bit task into L_1-bit task (Task 1) and L_2-bit task (Task 2)

10. Applying NOMA, U offloads Task 1 to AP_A and Task 2 to AP_B

11. **end if**

12. **exit**

In the next section, we will describe the particular math formulas for each phase.

• *EH Phase*

During this phase, the user U will collect energy from P using SC or MRC technique. For SC scheme, an antenna at U is selected for energy transmission to maximize the channel gain of the link from P to U. Thus, the index of the selected received antenna is given by:

$$\hat{n} = \arg \max_{1 \le n \le N} \left\{ |h_{0n}|^2 \right\}$$ (6.1)

For MRC technique, all of N antennas at user U are used for energy transmission.

The energy received at U during the time period $\tau_0 = \alpha T$ is:

$$E_h = \eta P_0 X \alpha T$$ (6.2)

where $0 < \eta \le 1$ energy conversion efficiency of the energy receiver, $0 < \alpha < 1$ time switching ratio, P_0 is the transmit power of power station P, and X denotes channel power gain of link P–U.

$$X = \begin{cases} \max_{1 \le n \le N} \left\{ |h_{0n}|^2 \right\} & ,SC \\ \sum_{n=1}^{N} |h_{0n}|^2 & ,MRC \end{cases}$$ (6.3)

The transmit power P_T is calculated as follows:

$$P_T = \frac{E_h}{(1-a)T - \tau} = \frac{\eta P_0 X \alpha T}{(1-a)T - \tau} = a P_0 X$$ (6.4)

where $a \triangleq \dfrac{\eta \alpha T}{(1-\alpha)T - \tau}$, $\tau = \max\{\tau_{21}, \tau_{22}\}$, τ_{2i} is the computation time for the L_i bit at AP_i, $i \in \{A, B\}$, τ_{2i} is defined as $\tau_{2i} = \dfrac{c_i L_i}{f_i}$, c_i is the number of CPU cycles needed to process the one input bit task at AP_i, and f_i is the operating frequency of AP_i.

• *Offloading Phase*

During the offloading phase, applying the NOMA technique, U transmits the superimposed signal as follows:

$$x = \sqrt{bP_T}\, s_1 + \sqrt{(1-b)P_T}\, s_2 \tag{6.5}$$

where s_1 and s_2 are L_1-bits task and L_2-bits task; b is the power allocation coefficient. Because AP_A has high-security requirements, so it is prioritized to allocate more power than AP_B, so b is chosen to satisfy the condition $0.5 < b \le 1$

Thus, the signal received at AP_A and AP_B from the nth antenna of U is given by

$$y_{ni} = \left(\sqrt{bP_T}\, s_1 + \sqrt{(1-b)P_T}\, s_2 \right) h_{ni} + \upsilon_{ni} \tag{6.6}$$

where υ_{ni} denotes the additive white Gaussian noise (AWGN) with zero mean and variance σ^2 at APs, $i \in \{1, 2\}$.

In case U uses TAS A scheme, an antenna of U, denoted as n_A, is selected for transmission to maximize the channel gain of link $U\text{-}AP_A$. Thus, the index of the selected received antennas is given by:

$$n_A = \arg\max_{1 \le n \le N} \left(|h_{n1}|^2 \right) \tag{6.7}$$

Then, the channel power gains for the selected transmit antenna of the links $U\text{-}AP_A$, $U\text{-}AP_B$ are:

$$(Y, Z) = \left(|h_{n_A 1}|^2, |h_{n_A 2}|^2 \right) \tag{6.8}$$

In case U uses TAS B scheme, an antenna of U, denoted as n_B, is selected for transmission to maximize the channel gain of link $U\text{-}AP_B$. Thus, the index of the selected received antennas is as

$$n_B = \arg\max_{1 \le n \le N} \left(|h_{n2}|^2 \right) \tag{6.9}$$

Then, the channel power gains for the selected transmit antenna of the links $U\text{-}AP_A$, $U\text{-}AP_B$ are:

$$(Y, Z) = \left(|h_{n_B 1}|^2, |h_{n_B 2}|^2 \right) \tag{6.10}$$

In case U uses the MRT scheme, all of N antennas of U are used in the communication process. Then, the channel power gain for the of the links U-AP_A, U-AP_B is given by:

$$(Y,Z) = \left(\sum_{n=1}^{N} \left(|h_{n1}|^2 \right), \sum_{n=1}^{N} \left(|h_{n2}|^2 \right) \right) \tag{6.11}$$

- *Computing Phase*

During this phase, the APs decode the received signal and compute the corresponding tasks. The instantaneous signal-to-interference-noise ratio (SINR) at AP_A for decoding s_1 is given by

$$\gamma_{AP_A}^{s1} = \frac{aP_0 XbY}{aP_0 X(1-b)Y + \sigma^2} = \frac{a\gamma_0 XY}{a(1-b)\gamma_0 XY + 1} \tag{6.12}$$

where $\gamma_0 \triangleq \dfrac{P_0}{\sigma^2}$ denotes the average transmit signal-to-noise ratio (SNR).

AP_B applies a successive interference cancellation (SIC) technique for decoding the signal s_1; then subtracting this factor from the received signal to obtain s_2. Hence, the instantaneous SINR at AP_B to decode s_1 is given by:

$$\gamma_{AP_B}^{s1} = \frac{a\gamma_0 XZ}{a(1-b)\gamma_0 XZ + 1} \tag{6.13}$$

And the instantaneous SNR at AP_B to decode s_2 is given by:

$$\gamma_{AP_B}^{s1} = \frac{aP_0 X(1-b)Z}{\sigma^2} = a(1-b)\gamma_0 XZ \tag{6.14}$$

The instantaneous channel capacity of link U-AP_A, and U-AP_B are given by, respectively:

$$C_1 = (1-\alpha)B\log_2 \left(1 + \gamma_{AP_A}^{s1} \right) \tag{6.15}$$

$$C_2 = (1-\alpha)B\log_2 \left(1 + \gamma_{AP_B}^{s2} \right) \tag{6.16}$$

where B is the channel bandwidth.

Let ε is the data allocation coefficient. Then $L_1 = \varepsilon L$ and $L_2 = (1-\varepsilon)L$. The computation time of tasks at the APs is rewritten to:

$$\tau_{21} = \frac{c_1 L_1}{f_1} = \frac{c_1 \varepsilon L}{f_1} \tag{6.17}$$

$$\tau_{22} = \frac{c_2 L_2}{f_2} = \frac{c_2(1-\varepsilon)L}{f_2} \tag{6.18}$$

• *Downloading Phase*

After the calculation is successful, the APs will feedback the calculation result to U in a short time.

Thus, depending on the diversity techniques selected at the EH phase and the Offloading phase, the system will be able to operate based on the four schemes summarized in Table 6.2.

We derive the probability density function (PDF), and cumulative distribution function (CDF) of the channel power gains to perform further analyses, i.e., X, Y, and Z. Because the channel gains follow the Nakagami distribution, for the SC schemes, the CDF of X is expressed as follows:

$$F_X(x) = \sum_{k=0}^{N} \sum_{\Delta_{scX}=k} \Phi_{scX} . x^{\varphi_{scX}} . \exp\left(\frac{-km_{PU}}{\lambda_{PU}} x \right) \tag{6.19}$$

$$\Phi_{scX} = \binom{N}{k} \binom{k}{\delta_0,...,\delta_{m_{PU}-1}} (-1)^k \left[\prod_{i=0}^{m_{PU}-1} \left(\frac{m_{PU}^i}{i! \lambda_{PU}^i} \right)^{\delta_i} \right], \Delta_{scX} = \sum_{i=0}^{m_{PU}-1} \delta_i, \varphi_{scX} = \sum_{i=0}^{m_{PU}-1} i\delta_i$$

$$f_X(x) = \sum_{k=1}^{N} \sum_{\Delta_{scX}=k} \Phi_{scX} \left(\varphi_{scX} . x^{\varphi_{scX}-1} . \exp\left(\frac{-km_{PU}}{\lambda_{PU}} x \right) + x^{\varphi_{scX}} . \exp\left(\frac{-km_{PU}}{\lambda_{PU}} x \right) . \left(\frac{-km_{PU}}{\lambda_{PU}} \right) \right)$$

$$\tag{6.20}$$

Because the channel gains follow the Nakagami distribution, for the TAS A schemes, the CDF of Y and Z are expressed as follows:

$$F_Y(x) = \sum_{l=0}^{N} \sum_{\Delta_{tasY}=l} \Phi_{tasY} . x^{\varphi_{tasY}} . \exp\left(\frac{-lm_{UA}}{\lambda_{UA}} x \right) \tag{6.21}$$

$$\Phi_{tasY} = \binom{N}{l}\binom{l}{\sigma_0,\ldots,\sigma_{m_{UA}-1}}(-1)^k\left[\prod_{j=0}^{m_{UA}-1}\left(\frac{m_{UA}^j}{j!\lambda_{PU}^j}\right)^{\sigma_j}\right], \Delta_{tasY} = \sum_{j=0}^{m_{UA}-1}\sigma_j, \varphi_{tasY} = \sum_{j=0}^{m_{UA}-1}j\delta_j$$

$$f_Y(x) = \sum_{l=0}^{N}\sum_{\Delta_{tasY}=l}\Phi_{tasY}\left(\varphi_{tasY}\cdot x^{\varphi_{tasY}-1}\cdot\exp\left(\frac{-lm_{UA}}{\lambda_{UA}}x\right)+x^{\varphi_{tasY}}\cdot\exp\left(\frac{-lm_{UA}}{\lambda_{UA}}x\right)\cdot\left(\frac{-lm_{UA}}{\lambda_{UA}}\right)\right)$$

$$(6.22)$$

$$F_Z(x) = 1-\exp\left(\frac{-m_{UB}}{\lambda_{UB}}x\right)\sum_{j=0}^{m_{UB}-1}\frac{1}{j!}\left(\frac{m_{UB}}{\lambda_{UB}}x\right)^j \qquad (6.23)$$

$$f_Z(x) = \frac{1}{(m_{UB}-1)!}\left(\frac{m_{UB}}{\lambda_{UB}}\right)^{m_{UB}}x^{m_{UB}-1}\cdot\exp\left(\frac{-m_{UB}}{\lambda_{UB}}x\right) \qquad (6.24)$$

Because the channel gains follow the Nakagami distribution, for the TAS B schemes, the CDF of Y and Z are expressed as follows:

$$F_Z(x) = \sum_{q=0}^{N}\sum_{\Delta_{tasZ}=q}\Phi_{tasZ}\cdot x^{\varphi_{tasZ}}\cdot\exp\left(\frac{-qm_{UB}}{\lambda_{UB}}x\right) \qquad (6.25)$$

$$\Phi_{tasZ} = \binom{N}{q}\binom{q}{\omega_0,\ldots,\omega_{m_{PU}-1}}(-1)^q\left[\prod_{p=0}^{m_{UB}-1}\left(\frac{m_{UB}^p}{p!\lambda_{UB}^p}\right)^{\omega_p}\right], \Delta_{tasZ} = \sum_{p=0}^{m_{UB}-1}\omega_p, \varphi_{tasZ} = \sum_{p=0}^{m_{UB}-1}p\delta_p$$

$$f_Z(x) = \sum_{p=0}^{N}\sum_{\Delta_{tasZ}=p}\Phi_{tasZ}\left(\varphi_{tasZ}\cdot x^{\varphi_{tasZ}-1}\cdot\exp\left(\frac{-pm_{UB}}{\lambda_{UB}}x\right)+x^{\varphi_{tasZ}}\cdot\exp\left(\frac{-m_{UB}}{\lambda_{UB}}x\right)\cdot\left(\frac{-qm_{UB}}{\lambda_{UB}}\right)\right)$$

$$(6.26)$$

$$F_Y(x) = 1-\exp\left(\frac{-m_{UA}}{\lambda_{UA}}x\right)\sum_{p=0}^{m_{UA}-1}\frac{1}{p!}\left(\frac{m_{UA}}{\lambda_{UA}}x\right)^p \qquad (6.27)$$

$$f_Y(x) = \frac{1}{(m_{UA}-1)!}\left(\frac{m_{UA}}{\lambda_{UA}}\right)^{m_{UA}}x^{m_{UA}-1}\cdot\exp\left(\frac{-m_{UA}}{\lambda_{UA}}x\right) \qquad (6.28)$$

Because the channel gains follow the Nakagami distribution, for the MRC/MRT schemes, the CDF of X, Y and Z are expressed as follows:

$$F_{gi}^{MRC/MRT}(x) = 1 - \sum_{l=0}^{m_i N-1} \frac{m_i^l}{l! \lambda_i^l} x^l \exp\left(-\frac{m_i}{\lambda_i}x\right) \qquad (6.29)$$

$$f_{gi}^{MRC/MRT}(x) = \frac{1}{(m_i N-1)!}\left(\frac{m_i}{\lambda_i}\right)^{m_i N} x^{m_i N-1} \exp\left(-\frac{m_i}{\lambda_i}x\right) \quad (6.30)$$

where $i \in \{1, 2, 3\}$.

6.3 Performance Analysis and Optimization

6.3.1 Performance Analysis

In this section, we investigate the RF EH NOMA-MEC system performance using a successful computation probability. The probability of successful computation Pr_s is defined as the probability that all tasks are completed by the maximum allowable delay time $T > 0$. In the proposed system, Pr_s is given by the formula:

$$Pr_s = Pr\left[\tau_1 + \tau \le (1-\alpha)\right]T \qquad (6.31)$$

Based on the APS algorithms, the SCP can be rewritten:

$$Pr_s = Pr[\tau_1 + \tau \le (1-\alpha)T] = \underbrace{Pr\left(Y > Z, C > \frac{L}{\Omega}\right)}_{P_1} + \underbrace{Pr\left(Y < Z, C_1 > \frac{L_1}{\Omega_1}, C_2 > \frac{L_2}{\Omega_2}\right)}_{P_2}$$

where $\Omega = (1-\alpha)T - \frac{c_1 L}{f_1}$, $\Omega_1 = (1-\alpha)T - \frac{c_1 L_1}{f_1}$, $\Omega_2 = (1-\alpha)T - \frac{c_2 L_2}{f_2}$ and $C = (1-\alpha)B\log_2(1+a\gamma_0 XY)$

To evaluate system performance, we present the following four lemmas.

Lemma 1: Under Nakagami-m fading, the SCP of the proposed system when using the SC-TAS A schemes is described in Eq. (6.32).

$$\Pr_s^{SC-TASA} = \begin{cases} P_{11}^{SC-TASA} & b < \rho \\ P_{11}^{SC-TASA} + P_{12}^{SC-TASA} & b > \rho \end{cases} \tag{6.32}$$

$$P_{11}^{SC-TASA} = \sum_{k=0}^{N}\sum_{\Delta_{scX}=k}^{N}\sum_{l=0}^{N}\sum_{\Delta_{tasY}=l} \Phi_{scX}.\Phi_{tasY}.\frac{\pi}{2M}\sum_{i=1}^{M}\theta_i^{\Lambda_{PU}-1}\sqrt{1-\alpha_i^2}\left[\Lambda_{UA}^{-\varphi_{tasY}}.\left(\varphi_{scX}\left(-\ln\theta_i\right)^{\varphi_{scX}-1} - \Lambda_{PU}\left(-\ln\theta_i\right)^{\varphi_{scX}}\right)\right.$$

$$*\left(\varphi_{tasY}.\Gamma\left(\varphi_{tasY},\frac{-\Lambda_{UA}\beta}{\ln\theta_i}\right) - \Gamma\left(\varphi_{tasY}+1,\frac{-\Lambda_{UA}\beta}{\ln\theta_i}\right)\right) - \sum_{j=0}^{m_{UB}-1}\frac{1}{j!}\left(\frac{m_{UB}}{\lambda_{UB}}\right)^j *\left(\varphi_{scX}\left(-\ln\theta_i\right)^{\varphi_{scX}-1} - \Lambda_{PU}\left(-\ln\theta_i\right)^{\varphi_{scX}}\right)$$

$$\left. *\left(\varphi_{tasY}.\Lambda_1^{-\varphi_{tasY}-j}.\Gamma\left(\varphi_{tasY}+j,\frac{-\Lambda_1\beta}{\ln\theta_i}\right) - \Lambda_{UA}.\Lambda_1^{-\varphi_{tasY}-j-1}.\Gamma\left(\varphi_{tasZ}+j+1,\frac{-\Lambda_1\beta}{\ln\theta_i}\right)\right)\right]$$

$$P_{12}^{SC-TASA} = \sum_{k=0}^{N}\sum_{\Delta_{scX}=k}^{N}\sum_{l=0}^{N}\sum_{\Delta_{tasY}=l} \Phi_{scX}.\Phi_{tasY}.\frac{\pi}{2M}\sum_{i=1}^{M}\theta_i^{\Lambda_{PU}-1}\sqrt{1-\alpha_i^2}\left[\left(\frac{1}{(m_{UB}-1)!}.\left(\frac{m_{UB}}{\lambda_{UB}}\right)^{m_{UB}}\right.\right.$$

$$*\left(\varphi_{scX}\left(-\ln\theta_i\right)^{\varphi_{scX}-1} - \Lambda_{PU}\left(-\ln\theta_i\right)^{\varphi_{scX}}\right).\Lambda_1^{-\varphi_{tasY}-m_{UB}}.\Gamma\left(\varphi_{tasY}+m_{UB},\frac{-\Lambda_1\beta_2}{\ln\theta_i}\right)$$

$$-(\beta_1)^{\varphi_{tasY}}.\frac{1}{(m_{UB}-1)!} * \exp\left(\frac{\Lambda_{UA}\beta_1}{\ln\theta_i}\right)\left(\varphi_{scX}\left(-\ln\theta_i\right)^{\varphi_{scX}-1-\varphi_{tasY}} - \Lambda_{PU}\left(-\ln\theta_i\right)^{\varphi_{scX}-\varphi_{tasY}}\right).\Gamma\left(m_{UB},\frac{-m_{UB}}{\lambda_{UB}}\frac{\beta_2}{\ln\theta_i}\right)\right]$$

where $\theta_i = \frac{\alpha_i+1}{2}, \alpha_i = \cos\left(\frac{2i-1}{2M}\pi\right), \Lambda_{PU} = \frac{km_{PU}}{\lambda_{PU}}, \Lambda_{UA} = \frac{lm_{UA}}{\lambda_{UA}},$

$\Lambda_1 = \frac{m_{UB}}{\lambda_{UB}} + \frac{lm_{UA}}{\lambda_{UA}}, \beta = \frac{2^{\frac{L}{(1-\alpha)B\Omega}}-1}{a\gamma_0}$

$\beta_1 = \dfrac{2^{\frac{L_1}{\Omega_1(1-\alpha)B}}-1}{\left(b-(1-b)*2^{\frac{L_1}{\Omega_1(1-\alpha)B}}-1\right)a\gamma_0}, \beta_2 = \dfrac{2^{\frac{L_2}{\Omega_2(1-\alpha)B}}-1}{(1-b)a\gamma_0}, \rho = 1 - \dfrac{1}{2^{\frac{L_1}{(1-\alpha)\Omega_1 B}}}$ and

M is the complexity-vs-accuracy trade-off coefficient.

Proof. See Appendix A.   ∎

Lemma 2: Under Nakagami-m fading, the SCP of the proposed system when using the SC-TAS B schemes is described in Eq. (6.33).

$$
\Pr_s^{SC-TASB} = \begin{cases} P_{11}^{SC-TASB} & b < \rho \\ P_{11}^{SC-TASB} + P_{12}^{SC-TASB} & b > \rho \end{cases} \tag{6.33}
$$

$$
P_{11}^{SC-TASB} = \sum_{k=0}^{N}\sum_{\Delta_{scX}=k}^{N}\sum_{q=0}^{N}\sum_{\Delta_{tasZ}=q} \Phi_{scX}.\Phi_{tasZ}.\frac{\pi}{2M}\sum_{i=1}^{M}\theta_i^{\Lambda_{PU}-1}\sqrt{1-\alpha_i^2}\left[\frac{1}{(m_{UA}-1)!}.\left(\frac{m_{UA}}{\lambda_{UA}}\right)^{m_{UA}}.\Lambda_2^{-\varphi_{tasZ}-m_{UA}}\right.
$$

$$
\star\left(\varphi_{scX}\left(-\ln\theta_i\right)^{\varphi_{scX}-1} - \Lambda_{PU}\left(-\ln\theta_i\right)^{\varphi_{scX}}\right).\Gamma\left(\varphi_{tasZ}+m_{UA},\frac{-\Lambda_2\beta}{\ln\theta_i}\right)\Bigg]
$$

$$
P_{12}^{SC-TASB} = \sum_{k=0}^{N}\sum_{\Delta_{scX}=k}^{N}\sum_{q=0}^{N}\sum_{\Delta_{tasZ}=q} \Phi_{scX}.\Phi_{tasZ}.\frac{\pi}{2M}\sum_{i=1}^{M}\theta_i^{\Lambda_{PU}-1}\sqrt{1-\alpha_i^2}\left[\sum_{p=0}^{m_{UA}-1}\frac{1}{p!}\left(\frac{m_{UA}}{\lambda_{UA}}\right)^p.\Lambda_{UB}^{-\varphi_{tasZ}}.\beta_1^p\right.
$$

$$
\star\exp\left(\frac{m_{UA}}{\lambda_{UA}}.\frac{\beta_1}{\ln\theta_i}\right).\left(\varphi_{scX}\left(-\ln\theta_i\right)^{\varphi_{scX}-1-p} - \Lambda_{PU}\left(-\ln\theta_i\right)^{\varphi_{scX}-p}\right).\left[\varphi_{tasZ}.\Gamma\left(\varphi_{tasZ},\frac{-\Lambda_{UB}\beta_2}{\ln\theta_i}\right)\right.
$$

$$
\left. -\Gamma\left(\varphi_{tasZ}+1,\frac{-\Lambda_{UB}\beta_2}{\ln\theta_i}\right)\right) - \left(\varphi_{scX}\left(-\ln\theta_i\right)^{\varphi_{scX}-1} - \Lambda_{PU}\left(-\ln\theta_i\right)^{\varphi_{scX}}\right)\star\left(\varphi_{tasZ}.\Lambda_2^{-\varphi_{tasZ}-p}.\Gamma\left(\varphi_{tasZ}+p,\frac{-\Lambda_2\beta_2}{\ln\theta_i}\right)\right.
$$

$$
\left.\left. -\Lambda_{UB}.\Lambda_2^{-\varphi_{tasZ}-p-1}.\Gamma\left(\varphi_{tasZ}+p+1,\frac{-\Lambda_2\beta_2}{\ln\theta_i}\right)\right)\right]
$$

where $\theta_i = \dfrac{\alpha_i+1}{2}, \alpha_i = \cos\left(\dfrac{2i-1}{2M}\pi\right), \Lambda_{PU} = \dfrac{km_{PU}}{\lambda_{PU}}, \Lambda_{UA} = \dfrac{lm_{UA}}{\lambda_{UA}}, \Lambda_2 = \dfrac{qm_{UB}}{\lambda_{UB}}$

$+\dfrac{m_{UA}}{\lambda_{UA}}, \beta = \dfrac{2^{\frac{L}{(1-\alpha)B\Omega}}-1}{a\gamma_0}$

$\beta_1 = \dfrac{2^{\frac{L_1}{\Omega_1(1-\alpha)B}}-1}{\left(b-(1-b)*2^{\frac{L_1}{\Omega_1(1-\alpha)B}}-1\right)a\gamma_0}, \beta_2 = \dfrac{2^{\frac{L_2}{\Omega_2(1-\alpha)B}}-1}{(1-b)a\gamma_0}, \rho = 1-\dfrac{1}{2^{\frac{L_1}{(1-\alpha)\Omega_1 B}}}$ and

M is the complexity-vs-accuracy trade-off coefficient.

Proof. The proof of Lemma 2 is similar to the proof of Lemma 1; see Appendix A.   ∎

Lemma 3: Under Nakagami-m fading, the SCP of the proposed system when using the SC-MRT schemes is described in Eq. (6.34).

$$\Pr_{i}{}^{SC-MRT} = \begin{cases} P_{11}^{SC-MRT} & b<\rho \\ P_{11}^{SC-MRT} + P_{11}^{SC-MRT} & b>\rho \end{cases} \tag{6.34}$$

$$P_{11}^{SC-MRT} = \sum_{k=0}^{N} \sum_{\Delta_{uX}=k} \Phi_{uX} \cdot \frac{\pi}{2M} \sum_{i=1}^{M} \theta_i^{\Lambda_{PU}-1} \sqrt{1-\alpha_i^2} \left[\frac{1}{(m_{UA}N-1)!} \cdot \left(\varphi_{uX} \left(-\ln\theta_i \right)^{\varphi_{uX}-1} - \Lambda_{PU} \left(-\ln\theta_i \right)^{\varphi_{uX}} \right) \right.$$

$$\left. {}^* \Gamma \left(m_{UA}N, \frac{-m_{UA}}{\lambda_{UA}} \frac{\beta}{\ln\theta_i} \right) - \sum_{w=0}^{m_{UB}N-1} \frac{1}{w!} \left(\frac{m_{UB}}{\lambda_{UB}} \right)^w \left(\frac{m_{UA}}{\lambda_{UA}} \right)^{m_{UA}N} \cdot \frac{1}{(m_{UA}N-1)!} \cdot \Lambda_3^{-m_{UA}N-w} \cdot \left(\varphi_{uX} \left(-\ln\theta_i \right)^{\varphi_{uX}-1} - \Lambda_{PU} \left(-\ln\theta_i \right)^{\varphi_{uX}} \right) \right.$$

$$\left. {}^* \Gamma \left(m_{UA}N+w, \frac{-\Lambda_3\beta}{\ln\theta_i} \right) \right]$$

$$P_{12}^{SC-MRT} = \sum_{k=0}^{N} \sum_{\Delta_{uX}=k} \Phi_{uX} \cdot \frac{\pi}{2M} \sum_{i=1}^{M} \theta_i^{\Lambda_{PU}-1} \sqrt{1-\alpha_i^2} \frac{1}{(m_{UA}N-1)!} \cdot \left[\left(\varphi_{uX} \left(-\ln\theta_i \right)^{\varphi_{uX}-1} - \Lambda_{PU} \left(-\ln\theta_i \right)^{\varphi_{uX}} \right) \cdot \Gamma \left(m_{UB}N, \frac{-m_{UB}}{\lambda_{UB}} \frac{\beta_2}{\ln\theta_i} \right) \right.$$

$$\left. + \frac{1}{w!} \left(\frac{m_{UA}}{\lambda_{UA}} \right)^w \cdot \frac{1}{(m_{UB}N-1)!} \left(\frac{m_{UB}}{\lambda_{UB}} \right)^{m_{UB}N} \cdot \Lambda_3^{-m_{UB}N-w} \left(\varphi_{uX} \left(-\ln\theta_i \right)^{\varphi_{uX}-1} - \Lambda_{PU} \left(-\ln\theta_i \right)^{\varphi_{uX}} \right) \cdot \Gamma \left(m_{UB}N+w, \frac{-\Lambda_3\beta_2}{\ln\theta_i} \right) \right.$$

$$\left. - \frac{1}{(m_{UB}N-1)!} \left[1 - \sum_{w=0}^{m_{UA}N-1} \frac{1}{w!} \left(\frac{m_{UA}}{\lambda_{UA}} \right)^w \cdot \left(-\ln\theta_i \right)^w \theta_i^{\frac{m_{UA}}{\lambda_{UA}}} \right] \left(\varphi_{uX} \left(-\ln\theta_i \right)^{\varphi_{uX}-1} - \Lambda_{PU} \left(-\ln\theta_i \right)^{\varphi_{uX}} \right) {}^* \Gamma \left(m_{UB}N, \frac{-m_{UB}}{\lambda_{UB}} \frac{\beta_2}{\ln\theta_i} \right) \right]$$

where $\theta_i = \dfrac{\alpha_i+1}{2}, \alpha_i = \cos\left(\dfrac{2i-1}{2M}\pi \right), \Lambda_{PU} = \dfrac{km_{PU}}{\lambda_{PU}}, \Lambda_{UA} = \dfrac{lm_{UA}}{\lambda_{UA}}, \Lambda_3 = \dfrac{m_{UA}}{\lambda_{UA}}$

$+\dfrac{m_{UB}}{\lambda_{UB}}, \beta = \dfrac{2^{\frac{L}{(1-\alpha)B\Omega}}-1}{a\gamma_0}$

$\beta_2 = \dfrac{2^{\frac{L_2}{\Omega_2(1-\alpha)B}}-1}{(1-b)a\gamma_0}, \rho = 1 - \dfrac{1}{2^{\frac{L_1}{(1-\alpha)\Omega_1 B}}}$ and M is the complexity-vs-accuracy trade-off coefficient.

Proof. The proof of Lemma 3 is similar to the proof of Lemma 1; see Appendix A.   ∎

Lemma 4: Under Nakagami-m fading, the SCP of the proposed system when using the MRC-MRT schemes is described in Eq. (6.35).

$$\Pr_{s}{}^{MRC-MRT} = \begin{cases} P_{11}^{MRC-MRT} & b<\rho \\ P_{11}^{MRC-MRT} + P_{11}^{MRC-MRT} & b>\rho \end{cases} \tag{6.35}$$

$$P_{11}^{MRC-MRT} = \frac{\pi}{2M} \sum_{i=1}^{M} \theta_i^{\frac{m_{PU}}{\lambda_{PU}}-1} \sqrt{1-\alpha_i^2} \left[\frac{1}{(m_{PU}N-1)!} \cdot \left(\frac{m_{PU}}{\lambda_{PU}} \right)^{m_{PU}N} \cdot \frac{1}{(m_{UA}N-1)!} \cdot \left(-\ln\theta_i \right)^{m_{PU}N-1} {}^* \Gamma \left(m_{UA}N, \frac{-m_{UA}}{\lambda_{UA}} \frac{\beta}{\ln\theta_i} \right) \right.$$

$$\left. - \sum_{w=0}^{m_{UB}N-1} \frac{1}{w!} \left(\frac{m_{UB}}{\lambda_{UB}} \right)^w \left(\frac{m_{PU}}{\lambda_{PU}} \right)^{m_{PU}N} \cdot \frac{1}{(m_{PU}N-1)!} \cdot \frac{1}{(m_{UA}N-1)!} \left(\frac{m_{UA}}{\lambda_{UA}} \right)^{m_{UA}N} \cdot \Lambda_3^{-m_{UA}N-w} \cdot \left(-\ln\theta_i \right)^{m_{PU}N-1} {}^* \Gamma \left(m_{UA}N+w, \frac{-\Lambda_3\beta}{\ln\theta_i} \right) \right.$$

$$P_{12}^{MRC-MRT} = \frac{\pi}{2M} \sum_{i=1}^{M} \theta_i^{\frac{m_{PU}}{\lambda_{PU}}-1} \sqrt{1-\alpha_i^2} \left[\frac{1}{(m_{PU}N-1)!} \cdot \left(\frac{m_{PU}}{\lambda_{PU}}\right)^{m_{UP}N} \cdot \frac{1}{(m_{UB}N-1)!} \cdot (-\ln\theta_i)^{m_{PU}N-1} * \Gamma\left(m_{UB}N, \frac{-m_{UB}}{\lambda_{UB}} \frac{\beta_2}{\ln\theta_i}\right) \right.$$

$$+ \frac{1}{(m_{PU}N-1)!} \left(\frac{m_{PU}}{\lambda_{PU}}\right)^{m_{UP}N} \cdot \frac{1}{w!}\left(\frac{m_{UA}}{\lambda_{UA}}\right)^w \cdot \left(\frac{m_{UB}}{\lambda_{UB}}\right)^{m_{UB}N} \cdot (-\ln\theta_i)^{m_{PU}N-1} * \Gamma\left(m_{UA}N+w, \frac{-\Lambda_3\beta}{\ln\theta_i}\right) - \frac{1}{(m_{UB}N-1)!}$$

$$* \left(\frac{m_{PU}}{\lambda_{PU}}\right)^{m_{UP}N} \left[1 - \sum_{w=0}^{m_{UA}N-1} \frac{1}{w!}\left(\frac{m_{UA}}{\lambda_{UA}}\right)^w \cdot (-\ln\theta_i)^w \cdot \theta_i^{\frac{m_{UA}}{\lambda_{UA}}} \right] \cdot (-\ln\theta_i)^{m_{PU}N-1} * \Gamma\left(m_{UB}N, \frac{-m_{UB}}{\lambda_{UB}} \frac{\beta_2}{\ln\theta_i}\right) \right)$$

where $\theta_i = \dfrac{\alpha_i+1}{2}, \alpha_i = \cos\left(\dfrac{2i-1}{2M}\pi\right), \Lambda_{PU} = \dfrac{km_{PU}}{\lambda_{PU}}, \Lambda_{UA} = \dfrac{lm_{UA}}{\lambda_{UA}}, \Lambda_3 = \dfrac{m_{UA}}{\lambda_{UA}}$

$+ \dfrac{m_{UB}}{\lambda_{UB}}, \beta = \dfrac{2^{\frac{L}{(1-\alpha)B\Omega}}-1}{a\gamma_0}$

$\beta_2 = \dfrac{2^{\frac{L_2}{\Omega_2(1-\alpha)B}}-1}{(1-b)a\gamma_0}, \rho = 1 - \dfrac{1}{2^{\frac{L_1}{(1-\alpha)\Omega_1 B}}}$ and M is the complexity-vs-accuracy

trade-off coefficient.

Proof. The proof of Lemma 4 is similar to the proof of Lemma 1; see Appendix A. ∎

6.3.2   Optimization

Based on the proposed model and the mathematical model presented above, the problem of optimizing the successful computation probability is called the maximal successful computation probability (MSCP) and states the following:

$$(MSCP): \quad \max_{\alpha,b,\varepsilon}(\Pr_s) \tag{6.36a}$$

$$subject \quad t_0 + t_1 + t_2 \leq T \tag{6.36b}$$

$$0 < \alpha \leq 1 \tag{6.36c}$$

$$0.5 < b \leq 1 \tag{6.36d}$$

$$0 < \varepsilon \leq 1 \tag{6.36e}$$

where constraint (6.36b) ensures all tasks are processed within the time delay. The constraint (6.36c) represents the time switching ratio. The constraint (6.36d) represents the power allocation coefficient using NOMA. The term (6.36e) representsthe data allocation coefficient of the user U.

To solve the MSCP problem, we proposed two algorithmsbased on evolution process, namely GA-MSCP and PSO-MSCP.

The genetic algorithm [33, 34], is a metaheuristic stochastic search algorithm based on natural selection and genetic mechanism in biology, suitable for solving the complicated optimization problem. The algorithm step of MSCP based on a genetic algorithm, namely GA-MSCP, is shown in Algorithm 2. Firstly, the initial population is produced. We assume that there is $nPop$ individual in the population, the time switching ratio, the power allocation coefficient, and the user U's data allocation coefficient are defined as individual chromosomes. The chromosome vector of the kth ($k = 1, 2, ..., nPop$) individual is as follow:

$$X_k = (\alpha_k, b_k, \varepsilon_k) \tag{6.37}$$

Algorithm 2: Genetic algorithm approach for maximal successful computation probability (GA-MSCP)

Input:
    Population size $nPop$, crossover rate d, mutation rate μ, the maximum evolutionary generation $MaxIt$

Output: $Bestsol(\alpha^*, b^*, \varepsilon^*)$

Procedure GA-MSCP()

1. In it the population with $nPop$ individuals with random genes (α, b, ε)
2. **while** $k < MaxIt$ **do**
3. Use (6.32)/(6.33)/(6.34)/(6.35) to evaluate the population.
4. Update $Bestsol(\alpha^*, b^*, \varepsilon^*)$
5. Parent Roullete-whellselection
6. Uniform crossover
7. Mutation
8. Select the best $nPop$ individuals in population.
9. $k = k + 1$
10. **end while**
11. Return $Bestsol(\alpha^*, b^*, \varepsilon^*)$

In this chapter, we deploy real coding GA, and the initial values of all the individual chromosomes are randomly generated under the constraint

(6.36c), (6.36d), and (6.36e). The fitness functions are (6.32)/(6.33)/(6.34)/(6.35) depend on the system selected scheme. Fitness value indicates how properly an individual adapts to the environment, which is used to determine outstanding individuals. Next, we analyze time complexity for Algorithm 2. Suppose the population size is $nPop$, the maximum evolutionary generation is $MaxIt$, the crossover and mutation rate are d and μ respectively, and the chromosome size is n. The time complexity of the overallGA-MSCP is:

$$T_{GA-MSCP} = MaxIt\left[\left(nPop.\log nPop\right)+d.\frac{nPop}{2}.n+\mu.nPop.n\right]$$

$$(6.37)$$

The second algorithm we propose is based on the Particle Swarm Optimization (PSO) algorithm [35], namely PSO-MSCP, described explicitly in Algorithm 3. Like GA, PSO updates generations and finds the optimal solution. However, it does not use the *crossover* or mutation operators. PSO is based on the idea of a swarm's prey in a multidimensional search space where particles know how far away from the food and keep the closest position to the food they have reached. The best way to find food is to follow the leaders' particle most relative to the food. In PSO, each solution is characterized by two parameters, the present position x_i and velocity v_i, which is called a particle. At the same time, each particle has a fitness value evaluated by a fitness function. In this problem, the fitness function is the SCP functions. At the beginning, the swarm position is randomly initialized. During the travel in search space, each particle is influenced by two pieces of information: the best experience of the particle ith ($pBest_i$) and the common best experience among the members of the swarm ($gBest$). The PSO particles will traverse the search space by following the ones with the best current solutions (maximum adaptability). Specifically, after each discrete interval, each element's velocity and position are updated according to the formulas [36].

$$v_i(t+1)=\chi\left(v_i(t)+e_1.r_1.\left(pBest_i(t)-x_i(t)\right)+e_2.r_2.\left(gBest(t)-x_i(t)\right)\right)$$
$$x_i(t+1)=x_i(t)+v_i(t+1)$$

$$(6.38)$$

where $\chi = \dfrac{2}{\left|2-\phi-\sqrt{\phi^2-4\phi}\right|}$ is constriction factor with $\phi = \phi_1 + \phi_2 > 4$, r_1 and r_2 are random number follow Uniform distribution with a range of 0 to 1; $r_1, r_2 \sim U(0,1)$, $e_1 = \chi\phi_1, e_2 = \chi\phi_2$ are two acceleration coefficients, are respectively cognitive and social characteristics.

Algorithm 3: Particle Swarm Optimization algorithm approach for maximal successful computation probability (PSO-MSCP)

Input:
      Population size *nPop*, the maximum
      evolutionary generation *MaxIt*

Output: *Bestsol($\alpha^*,b^*,\varepsilon^*$)*

Procedure PSO-MSCP()

---

1. Init the Swarm particles *nPop* with random position x_i *(α,b,ε)*, zero velocity v_i, and acceleration coefficients e_1, e_2

2. **while** $k <$ *MaxIt* **do**

3.   **for i** = *1:nPop*

4.   Update x_i and v_i using *(6.38)*

5. Use*(6.32)/(6.33)/(6.34)/(6.35)* to evaluate the particles.

6. **if** (fitness value *<pBest*) **then** Update*Bestsol($\alpha^·,b^·,\varepsilon^·$)* and *BestCost*

7. **if** *(BestCost<gBest)* **then** Update *gBest*

8.   **end if**

9.   **end if**

10.   **end for**

11. $k = k +1$

12. **end while**

13. Return *Bestsol($\alpha^*,b^*,\varepsilon^*$)*

Quickly see the time complexity of PSO-MSCP is:

$$T_{PSO-MSCP} = MaxIt.nPop \tag{6.39}$$

6.4   Numerical Results and Discussion

In this section, we present numerical results on the SCP. A Monte Carlo simulation is used to validate the analysis results. Unless differently declared, the parameters needed throughout the simulation and analysis are shown in Table 6.3.

The impacts of the time switching ratio on the SCP are depicted as Figure 6.3 with $\gamma_0 = 10$ dB, $\varepsilon = 0.6$, $b = 0.8$ for different schemes. From this

Table 6.3 Typical values of the parameters in the simulation.

Parameters	Notation	Typical values
Number of antennas of user	N	2
Nakagami shape factor	m_{PU}, m_{UA}, m_{UB}	3, 2, 1
Average transmit SNR	γ_0	0–30 dB
Energy conversion efficiency	η	0.75
CPU-cycle frequency of AP_A	f_1	5 GHz
CPU-cycle frequency of AP_B	f_2	5 GHz
Number of CPU cycles needed to compute one input bit	c_1, c_2	10
The complexity-vs-accuracy trade-off coefficient	M	100
Channel bandwidth	B	100 MHz
The threshold of latency	T	0.1 s
The length of task	L	80 Mb
The population size	$nPop$	50
Crossover rate	d	0.98
Mutation rate	μ	0.002
The maximum evolutionary generation	$MaxIt$	500
	ϕ_1, ϕ_2	2.05

figure, we can observe that when α increases from 0 to α^*, SCP improves; when α continuously increases from α^* to 1, SCP degrades. It can be demonstrated that when α increases from 0 to α^*, the more harvested energy, the better SCP. However, if α continuously increases, the remaining time for transmission and computation is more concise; this causes SCP to degrade.

The impacts of power allocation coefficient on the SCP are depicted as Figure 6.4 with $\gamma_0 = 10$ dB, $\varepsilon = 0.6$, $\alpha = 0.2$. We observed that as b gradually increased from 0.5 to 1, SCP gradually increased to a peak, then gradually decreased. It could be explained when the energy allocated entirely for

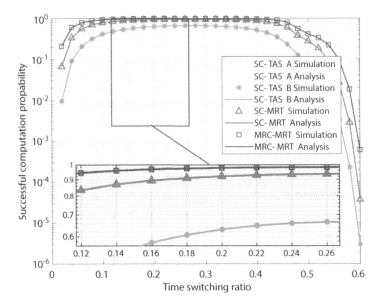

Figure 6.3 Impacts of the time switching ratio on SCP of different schemes.

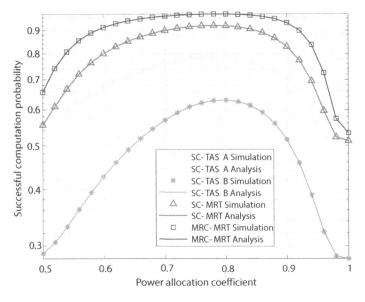

Figure 6.4 Impacts of the power allocation coefficient on SCP of different schemes.

sending information to AP_1 would result in no energy sending the information to AP_2, thereby depleting the system's SCP. There will inevitably exist a b^* that is the optimal value that maximizes the SCP.

We can conclude that the system using the MRC-MRT scheme brings the highest efficiency through the above surveys, while SC-TAS B offers the lowest system performance.

Figure 6.5 depicts the convergence of the GA-MSCP and PSO-MSCP algorithms. To make it easier to observe, we use the false computation probability (FCP) function as a fitness function. FCP is the probability U computed unsuccessfully and is defined as SCP's complement. In the scenario with the same $nPop$ parameter, PSO-MSCP can converge faster than GA-MSCP while also providing better system performance improvement.

Figure 6.6 depicts the SCP comparison between the system case using GA-MSCP, PSO-MSCP and the non-optimal case. The results show that the system performance is improved when using GA-MSCP or PSO-MSCP.

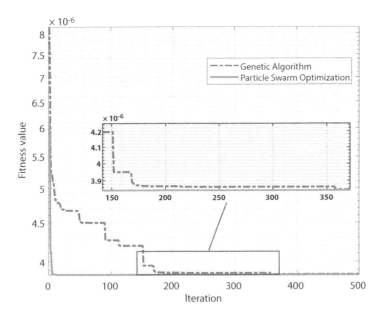

Figure 6.5 The convergence of optimization algorithms.

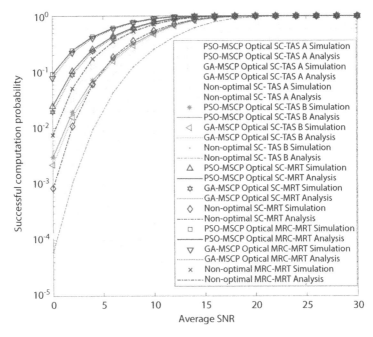

Figure 6.6 SCP comparison of different methods with/without optimal algorithms.

6.5   Conclusion

In this chapter, we have studied the RF energy harvesting NOMA mobile edge computing network. We proposed four schemes APS, namely SC-TAS A, SC-TAS B, SC-MRT, and MRC-MRT based on antenna and access point selection process. We derived the closed-form expressions of successful computation probability for these corresponding schemes using Gaussian–Chebyshev quadrature method. Moreover, we proposed the optimization algorithm GA-MSCP and PSO-MSCP to obtain the optimal performance of considered system. Finally, the numerical results have been provided to look insightthe system operation.

Appendix A

In this section, we prove the Lemma 1

$$P_1 = \Pr\left(Y > Z, (1-\alpha)B\log_2\left(a\gamma_0 XY + 1\right) > \frac{L}{\Omega}\right)$$

$$= \Pr\left(Y > Z, Y > \frac{\beta}{X}\right)$$

$$= \int_0^\infty \int_{\beta/y}^\infty F_Z(x).f_Y(x).f_X(y)\,dx\,dy$$

$$= \int_0^\infty \int_{\beta/y}^\infty f_X(y)*\left[1 - \exp\left(\frac{-m_{UB}}{\lambda_{UB}}x\right)\sum_{j=0}^{m_{UB}-1}\frac{1}{j!}\left(\frac{m_{UB}}{\lambda_{UB}}x\right)^j\right]$$

$$*\left[\sum_{l=1}^N\sum_{\Delta_{tasY}=l}\Phi_{tasY}.\exp\left(\frac{-lm_{UA}}{\lambda_{UA}}x\right)\left(\varphi_{tasY}.x^{\varphi_{tasY}-1}+\left(\frac{-lm_{UA}}{\lambda_{UA}}\right)x^{\varphi_{tasY}}\right)\right]dxdy$$

$$\text{(A.1)}$$

Let $\Lambda_A = \dfrac{lm_{UA}}{\lambda_{UA}}$, the formula (A.1) is as follow:

$$P_1 = \underbrace{\int_0^\infty \int_{\beta/y}^\infty f_X(y)*\left[\sum_{l=1}^N\sum_{\Delta_{tasY}=l}\Phi_{tasY}.\exp(-\Lambda_A x)\left(x^{\varphi_{tasY}-1}.\varphi_{tasY}-\Lambda_A x^{\varphi_{tasY}}\right)\right]dx\,dy}_{P_{11}}$$

$$+\underbrace{\int_0^\infty \int_{\beta/y}^\infty f_X(y)*\left[\sum_{l=1}^N\sum_{\Delta_{tasY}=l}\sum_{j=0}^{m_{UB}-1}\Phi_{tasY}.\exp\left(\left(-\Lambda_A-\frac{m_{UB}}{\lambda_{UB}}\right)x\right)\frac{1}{j!}\left(\frac{m_{UB}}{\lambda_{UB}}\right)^j\left(x^{\varphi_{tasY}-1+j}.\varphi_{tasY}-\Lambda_A.x^{\varphi_{tasY}+j}\right)\right]dx\,dy}_{P_{12}}$$

$$\text{(A.2)}$$

Let $\Lambda_{PU} = \dfrac{km_{PU}}{\lambda_{PU}}$ and substitute (6.20) into P_{11}, we have:

$$P_{11} = \overset{a}{\int_0^\infty} \sum_{k=1}^{N} \sum_{\Delta_{scX}=k} \Phi_{scX} \cdot \exp\left(-\Lambda_{PU}\, y\right)\left(\varphi_{scX} \cdot y^{\varphi_{scX}-1} + y^{\varphi_{scX}}\left(-\Lambda_{PU}\right)\right)$$

$$* \sum_{l=1}^{N} \sum_{\Delta_{tasY}=l} \Phi_{tasY}\left(\varphi_{tasY} \cdot \Lambda_{UA}^{-\varphi_{tasY}} \cdot \Gamma\left(\varphi_{tasY}, \frac{\Lambda_{UA}\beta}{y}\right) - \Lambda_{UA} \cdot \Lambda_{UA}^{-\varphi_{tasY}-1} \cdot \Gamma\left(\varphi_{tasY}+1, \frac{\Lambda_{UA}\beta}{y}\right)\right)\right)dy$$

$$\text{(A.3)}$$

Note that step (a) is obtained by the aid of [37, Equation (3.351.2)]

Let $\displaystyle\sum_{k=1}^{N}\sum_{\Delta_{scX}=k}\sum_{l=1}^{N}\sum_{\Delta_{tasY}=l} \triangleq \Sigma_{XY}$ and $\Phi_{scX}\Phi_{tasY} = \Phi_{XY}, z = \exp(-y),$ (A.3) is rewritten:

$$P_{11} = \sum_{XY} \Phi_{XY}\Lambda_{UA}^{-\varphi_{tasY}} \int_0^1 z^{\Lambda_{PU}}\left(\varphi_{scX}\cdot(-\ln z)^{\varphi_{scX}-1} - \Lambda_{PU}\cdot(-\ln z)^{\varphi_{scX}}\right)$$

$$* \left(\varphi_{tasY}\cdot\Gamma\left(\varphi_{tasY}, \frac{-\Lambda_{UA}\beta}{\ln z}\right) - \Gamma\left(\varphi_{tasY}+1, \frac{-\Lambda_{UA}\beta}{\ln z}\right)\right)\left(\frac{1}{z}\right)dz \qquad \text{(A.4)}$$

Using Gaussian–Chebyshev quadrature method, we have:

$$P_{11} = \frac{\pi}{2M}\sum_{XY} \Phi_{XY}\cdot\Lambda_{UA}^{-\varphi_{tasY}}\cdot\sum_{i=1}^{M}\theta_i^{\Lambda_{PU}-1}\cdot\sqrt{1-\alpha_i^2}\cdot\left(\varphi_{scX}\left(-\ln\theta_i\right)^{\varphi_{scX}-1} - \Lambda_{PU}\left(-\ln\theta_i\right)^{\varphi_{scX}}\right)$$

$$* \left(\varphi_{tasY}\cdot\Gamma\left(\varphi_{tasY}, \frac{-\Lambda_{UA}\beta}{\ln\theta_i}\right) - \Gamma\left(\varphi_{tasY}+1, \frac{-\Lambda_{UA}\beta}{\ln\theta_i}\right)\right) \qquad \text{(A.5)}$$

where $\theta_i = \dfrac{\alpha_i+1}{2}, \alpha_i = \cos\left(\dfrac{2i-1}{2M}\pi\right)$

Next, we introduce the way to obtain P_{12}. Let $\Lambda_1 = \dfrac{lm_{UA}}{\lambda_{UA}} + \dfrac{m_{UB}}{\lambda_{UB}}$, the P_{12} is as follow:

$$P_{12} = \int_0^\infty \int_{\beta/y}^\infty f_X(y) * \left[\sum_{l=1}^N \sum_{\Delta_{tasY}=l} \sum_{j=0}^{m_{UB}-1} \Phi_{tasY} . \exp(-\Lambda_1 x) \frac{1}{j!} \left(\frac{m_{UB}}{\lambda_{UB}} \right)^j \left(x^{\varphi_{tasY}-1+j} . \varphi_{tasY} - \Lambda_{UA} . x^{\varphi_{tasY}+j} \right) \right] dx\, dy$$

$$= \int_0^b \sum_{k=0}^N \sum_{\Delta_{tasX}=k} \Phi_{scX} . \exp(-\Lambda_{PU}\, y) \left(\varphi_{scX} . y^{\varphi_{scX}-1} - \Lambda_{PU} . y^{\varphi_{scX}} \right) * \sum_{l=1}^N \sum_{\Delta_{tasY}=l} \sum_{j=0}^{m_{UB}-1} \Phi_{tasY} . \frac{1}{j!} \left(\frac{m_{UB}}{\lambda_{UB}} \right)^j$$

$$* \left[\varphi_{tasY} . \Lambda_1^{-\varphi_{tasY}-j} . \Gamma\left(\varphi_{tasY}+j, \frac{\Lambda_1\beta}{y} \right) - \Lambda_{UA} . \Lambda_1^{-\varphi_{tasY}-j-1} . \Gamma\left(\varphi_{tasY}+j+1, \frac{\Lambda_1\beta}{y} \right) \right] dy$$

$$= \frac{\pi}{2M} \sum_{XY} \Phi_{XY} \Lambda_{UA}^{-\varphi_{tasY}} \sum_{j=0}^{m_{UB}-1} \frac{1}{j!} \left(\frac{m_{UB}}{\lambda_{UB}} \right)^j \sum_{i=1}^M \theta_i^{\Lambda_{PU}-1} . \sqrt{1-\alpha_i^2}$$

$$* \left(\varphi_{scX} \left(-\ln\theta_i \right)^{\varphi_{scX}-1} - \Lambda_{PU} \left(-\ln\theta_i \right)^{\varphi_{scX}} \right)$$

$$* \left[\varphi_{tasY} \Lambda_1^{-\varphi_{tasY}-j} . \Gamma\left(\varphi_{tasY}+j, \frac{-\Lambda_1\beta}{\ln\theta_i} \right) \right] - \Lambda_{UA} . \Lambda_1^{-\varphi_{tasY}-j-1} . \Gamma\left(\varphi_{tasY}+j+1, \frac{-\Lambda_1\beta}{\ln\theta_i} \right) \tag{A.6}$$

Note that step (b) is obtained by the aid of [37, Equation (3.351.2)] and step (c) is obtained by using Gaussian–Chebyshev quadrature method. In the next step, we describe the step to obtain P_2.

$$P_2 = \Pr\left(Y < Z, C_1 > \frac{L_1}{\Omega_1}, C_2 > \frac{L_2}{\Omega_2} \right)$$

$$= \Pr\left(Y < Z, Y \left[a\gamma_0 X \left(b - (1-b) \left(2^{\frac{L_1}{(1-\alpha)B\Omega_1}} - 1 \right) \right) \right] \geq 2^{\frac{L_1}{(1-\alpha)B\Omega_1}} - 1, a(1-b)\gamma_0 XZ \geq 2^{\frac{L_2}{(1-\alpha)B\Omega_2}} - 1 \right)$$

$$= \begin{cases} 0 & , b \leq 1 - \dfrac{1}{\dfrac{2^{\frac{L_1}{(1-\alpha)B\Omega_1}}}{\rho}} \\ \\ \Pr\left(Y < Z, Y > \dfrac{\beta_1}{X}, Z > \dfrac{\beta_2}{X} \right) & , b > \rho \end{cases} \tag{A.7}$$

We focus the condition $b > p$

$$P_2 = \int_0^\infty \int_{\beta_2/y}^\infty \left[F_Y(x) - F_Y\left(\frac{\beta_1}{y}\right) \right] f_Z(x) f_X(y) dx dx$$

$$= \int_0^\infty f_X(y) \left[\underbrace{\int_{\beta_2/y}^\infty F_Y(x).f_Z(x)dx}_{P_{21}} - \underbrace{\int_{\beta_2/y}^\infty F_Y\left(\frac{\beta_1}{y}\right).f_Z(x)dx}_{P_{22}} \right] dy$$

Next, we calculate P_{21}

$$P_{21} = \int_0^\infty \sum_{k=1}^N \sum_{\Delta_{scX}=k} \Phi_{scX}.\exp\left(-\Lambda_{PU}y\right)\left(\varphi_{scX}.y^{\varphi_{scX}-1} - \Lambda_{PU}.y^{\varphi_{scX}}\right)$$

$$* \int_{\beta_2/y}^\infty \sum_{l=1}^N \sum_{\Delta_{tasY}=l} \Phi_{tasY}.x^{\varphi_{tasY}}.\exp\left(-\Lambda_{UA}x\right)* \frac{1}{(m_{UB}-1)!}\left(\frac{m_{UB}}{\lambda_{UB}}\right)^{m_{UB}}.x^{m_{UB}-1}.\exp\left(-\frac{m_{UB}}{\lambda_{UB}}x\right) dx dy$$

$$= \frac{\pi}{2M}\sum_{XY} \Phi_{XY} \frac{1}{(m_{UB}-1)!}\left(\frac{m_{UB}}{\lambda_{UB}}\right)^{m_{UB}}.\Lambda_1^{-\varphi_{tasY}-m_{UB}} \sum_{i=1}^M \theta_i^{\Lambda_{PU}-1}.\sqrt{1-\alpha_i^2}$$

$$* \left(\varphi_{scX}\left(-\ln\theta_i\right)^{\varphi_{scX}-1} - \Lambda_{PU}\left(-\ln\theta_i\right)^{\varphi_{scX}}\right) * \Gamma\left(\varphi_{tasY}+m_{UB}, \frac{-\Lambda_1\beta_2}{\ln\theta_i}\right) \tag{A.8}$$

Using the similar method, we easily find P_{22}

$$P_{22} = \frac{\pi}{2M}.\sum_{XY} \Phi_{XY}.\frac{1}{(m_{UB}-1)!}.\beta_1^{\varphi_{tasY}}.\sum_{i=1}^M \theta_i^{\Lambda_{PU}-1}.\sqrt{1-\alpha_i^2}.\exp\left(\frac{\Lambda_{UA}\beta_1}{\ln\theta_i}\right)$$

$$* \left(\varphi_{scX}.\left(-\ln\theta_i\right)^{\varphi_{scX}-1-\varphi_{tasY}} - \Lambda_{PU}.\left(-\ln\theta_i\right)^{\varphi_{scX}-\varphi_{tasY}}\right) * \Gamma\left(m_{UB}, \frac{m_{UB}}{\lambda_{UB}}.\frac{-\beta_2}{\ln\theta_i}\right) \tag{A.9}$$

This ends our proof.

References

1. Li, S., Da, Xu, L., Zhao, S., 5G Internet of Things: A survey. *J. Ind. Inf. Integr.*, *10*, 1–9, 2018. https://doi.org/10.1016/j.jii.2018.01.005.

2. Akpakwu, G.A., Silva, B.J., Hancke, G.P., Abu-Mahfouz, A.M., A survey on 5G networks for the Internet of Things: Communication technologies and challenges. *IEEE Access*, *6*, 3619–3647, 2017. https://doi.org/10.1109/ACCESS.2017.2779844.

3. Pham, Q.V., Fang, F., Ha, V.N., Piran, M.J., Le, M., Le, L.B., Ding, Z., A survey of multi-access edge computing in 5G and beyond: Fundamentals, technology integration, and state-of-the-art. *IEEE Access*, *8*, 116974–117017, 2020. https://doi.org/10.1109/ACCESS.2020.3001277.

4. Li, Z., Zhou, X., Qin, Y., A survey of mobile edge computing in the industrial Internet, in: *2019 7th International Conference on Information, Communication and Networks (ICICN)*, IEEE, pp. 94–98, 2019, April, https://doi.org/10.1109/ICICN.2019.8834959.

5. Mao, Y., You, C., Zhang, J., Huang, K., Letaief, K.B., A survey on mobile edge computing: The communication perspective. *IEEE Commun. Surv. Tut.*, *19*, 4, 2322–2358, 2017. https://doi.org/10.1109/COMST.2017.2745201.

6. Li, B., Fei, Z., Shen, J., Jiang, X., Zhong, X., Dynamic offloading for energy harvesting mobile edge computing: Architecture, case studies, and future directions. *IEEE Access*, *7*, 79877–79886, 2019. https://doi.org/10.1109/ACCESS.2019.2922362.

7. Zhang, H., Chen, Z., Wu, J., Deng, Y., Xiao, Y., Liu, K., Li, M., Energy-efficient online resource management and allocation optimization in multi-user multi-task mobile-edge computing systems with hybrid energy harvesting. *Sensors*, *18*, 9, 3140, 2018. https://doi.org/10.3390/s18093140.

8. Min, M., Wan, X., Xiao, L., Chen, Y., Xia, M., Wu, D., Dai, H., Learning-based privacy-aware offloading for healthcare IoT with energy harvesting. *IEEE Internet Things J.*, *6*, 3, 4307–4316, 2018. https://doi.org/10.1109/JIOT.2018.2875926.

9. Chen, W., Wang, D., Li, K., Multi-user multi-task computation offloading in green mobile edge cloud computing. *IEEE T. Serv. Comput.*, *12*, 5, 726–738, Sep. 2019. https://doi.org/10.1109/TSC.2018.2826544.

10. Zhang, G., Zhang, W., Cao, Y., Li, D., Wang, L., Energy-delay tradeoff for dynamic offloading in mobile-edge computing system with energy harvesting devices. *IEEE Trans. Industr. Inform.*, *14*, 10, 4642–4655, Oct. 2018. https://doi.org/10.1109/TII.2018.2843365.

11. Ulukus, S., Yener, A., Erkip, E., Simeone, O., Zorzi, M., Grover, P., Huang, K., Energy harvesting wireless communications: A review of recent advances. *IEEE J. Sel. Areas Commun.*, *33*, 3, 360–381, Mar. 2015. https://doi.org/10.1109/JSAC.2015.2391531.

12. Wu, H., Chen, L., Shen, C., Wen, W., Xu, J., Online geographical load balancing for energy-harvesting mobile edge computing, in: *2018 IEEE*

International Conference on Communications (ICC), Kansas City, MO, USA, May 2018, https://doi.org/10.1109/ICC.2018.8422299.

13. López, O.L.A., Alves, H., Latva-aho, M., Joint power control and rate allocation enabling ultra-reliability and energy efficiency in SIMO wireless networks. *IEEE Trans. Commun.*, *67*, 8, 5768–5782, 2019. https://doi.org/10.1109/TCOMM.2019.2914682.

14. Nguyen, T.N., Tran, P.T., Voznak, M., Wireless energy harvesting meets receiver diversity: A successful approach for two-way half-duplex relay networks over block Rayleigh fading channel. *Comput. Netw.*, 172, 107176, 2020. https://doi.org/10.1016/j.comnet.2020.107176.

15. Truong, V.T., Vo, M.T., Lee, Y., Ha, D.B., Amplify-and-Forward Relay Transmission in Uplink Non-Orthogonal Multiple Access Networks, in: *2019 6th NAFOSTED Conference on Information and Computer Science (NICS)*, IEEE, pp. 1–6, 2019, December, https://doi.org/10.1109/NICS48868.2019.9023818.

16. Truong, V.-T., Ha, D.-B., Lee, Y., Nguyen, A.-N., On Performance of Cooperative Transmission in Uplink Non-Orthogonal Multiple Access Wireless Sensor Networks. *2020 4th International Conference on Recent Advances in Signal Processing, Telecommunications & Computing (SigTelCom)*, Hanoi, Vietnam, pp. 56–60, 2020, https://doi.org/10.1109/SigTelCom49868.2020.9199063.

17. Zeng, M., Du, R., Fodor, V., Fischione, C., Computation Rate Maximization for Wireless Powered Mobile Edge Computing with NOMA, in: *2019 IEEE 20th International Symposium on "A World of Wireless, Mobile and Multimedia Networks" (WoWMoM)*, IEEE, pp. 1–9, 2019, June, https://doi.org/10.1109/TWC.2018.2821664.

18. Zhou, F., Wu, Y., Hu, R.Q., Qian, Y., Computation efficiency in a wireless-powered mobile edge computing network with NOMA, in: *ICC 2019-2019 IEEE International Conference on Communications (ICC)*, IEEE, pp. 1–7, 2019, May, https://doi.org/10.1109/ICC.2019.8761172.

19. Hoang, T.M., Van Son, V., Dinh, N.C., Hiep, P.T., Optimizing duration of energy harvesting for downlink NOMA full-duplex over Nakagami-m fading channel. *AEU-Int. J. Electron. Commun.*, *95*, 199–206, 2018. https://doi.org/10.1016/j.aeue.2018.08.020.

20. Rauniyar, A., Engelstad, P., Østerbø, O.N., RF energy harvesting and information transmission based on NOMA for wireless powered IoT relay systems. *Sensors*, 18, 10, 3254, 2018. https://doi.org/10.3390/s18103254.

21. Wang, F., Xu, J., Ding, Z., Optimized multiuser computation offloading with multi-antenna NOMA, in: *2017 IEEE Globecom Workshops (GC Wkshps)*, IEEE, pp. 1–7, 2017, December, https://doi.org/10.1109/GLOCOMW.2017.8269088.

22. Yang, Z., Pan, C., Hou, J., Shikh-Bahaei, M., Efficient resource allocation for mobile-edge computing networks with NOMA: Completion time and energy

minimization. *IEEE Trans. Commun.*, *67*, 11, 7771–7784, 2019. https://doi. org/10.1109/TCOMM.2019.2935717.

23. Zhou, F., Wu, Y., Sun, H., Chu, Z., UAV-enabled mobile edge computing: Offloading optimization and trajectory design, in: *2018 IEEE International Conference on Communications (ICC)*, IEEE, pp. 1–6, 2018, May, https://doi. org/10.1109/ICC.2018.8422277.

24. Nguyen Kieu, T., Tran, D.D., Ha, D.B., Voznak, M., On secure QoS-based NOMA networks with multiple antennas and eavesdroppers over Nakagami-m fading. *IETE J. Res.*, 1–13, 2019. https://doi.org/10.1080/03772 063.2019.1610088.

25. Garcia, C.E., Camana, M.R., Koo, I., Joint Beamforming and Artificial Noise Optimization for Secure Transmissions in MISO-NOMA Cognitive Radio System with SWIPT. *Electronics*, *9*, 11, 1948, 2020. https://doi.org/10.3390/ electronics9111948.

26. Fu, Y., Zhang, M., Salaun, L., Sung, C.W., Chen, C.S., Zero-forcing Oriented Power Minimization for Multi-cell MISO-NOMA Systems: A Joint User Grouping, Beamforming and Power Control Perspective. *IEEE J. Sel. Areas Commun.*, 38, 8, 1925–1940, 2020. https://doi.org/10.1109/ JSAC.2020.3000825.

27. Ha, D.B., Truong, V.T., Ha, D.H., A Novel Secure Protocol for Mobile Edge Computing Network Applied Downlink NOMA, in: *International Conference on Industrial Networks and Intelligent Systems*, Springer, Cham, pp. 324–336, 2020, August, https://doi.org/10.1007/978-3-030-63083-6_25.

28. Ding, Z., Ng, D.W.K., Schober, R., Poor, H.V., Delay minimization for NOMA-MEC offloading. *IEEE Signal Process. Lett.*, *25*, 12, 1875–1879, 2018. https://doi.org/10.1109/LSP.2018.2876019.

29. Ye, Y., Lu, G., Hu, R.Q., Shi, L., On the performance and optimization for MEC networks using uplink NOMA, in: *2019 IEEE International Conference on Communications Workshops (ICC Workshops)*, IEEE, pp. 1–6, 2019, https://doi.org/10.1109/ICCW.2019.8756980.

30. Ye, Y., Hu, R.Q., Lu, G., Shi, L., Enhance latency-constrained computation in MEC networks using uplink NOMA. *IEEE Trans. Commun.*, *68*, 4, 2409–2425, 2020. DOI: 10.1109/TCOMM.2020.2969666.

31. Chen, A., Yang, Z., Lyu, B., Xu, B., System Delay Minimization for NOMA-Based Cognitive Mobile Edge Computing. *IEEE Access*, 8, 62228–62237, 2020. https://doi.org/10.1109/ACCESS.2020.2984634.

32. Wang, Y., Sheng, M., Wang, X., Wang, L., Li, J., Mobile-edge comput-ing: Partial computation offloading using dynamic voltage scaling. *IEEE Trans. Commun.*, 64, 10, 4268–4282, 2016. https://doi.org/10.1109/ TCOMM.2016.2599530.

33. Al-Habob, A.A., Dobre, O.A., Armada, A.G., Muhaidat, S., Task Scheduling for Mobile Edge Computing Using Genetic Algorithm and Conflict Graphs.

IEEE Trans. Veh. Technol., 69, 8, 8805–8819, 2020. https://doi.org/10.1109/TVT.2020.2995146.

34. Li, Z. and Zhu, Q., Genetic Algorithm-Based Optimization of Offloading and Resource Allocation in Mobile-Edge Computing. *Information, 11*, 2, 83, 2020. https://doi.org/10.3390/info11020083.

35. Eberhart, R. and Kennedy, J., A new optimizer using particle swarm theory. *MHS'95. Proceedings of the Sixth International Symposium on Micro Machine and Human Science*, Nagoya, Japan, pp. 39–43, 1995, https://doi.org/10.1109/MHS.1995.494215.

36. Clerc, M. and Kennedy, J., The particle swarm—Explosion, stability, and convergence in a multidimensional complex space. *IEEE Trans. Evol. Comput.*, 6, 58–73, 2002. https://doi.org/https://doi.org/10.1109/4235.985692.

37. Gradshteyn, I. and Ryzhik, I., *Table of Integrals, Series, and Products*, D. Zwillinger, (Ed.), Elsevier Academic Press, 2007.

7

Performance and Stability Analysis of CNTFET SRAM Cell Topologies for Ultra-Low Power Applications

Hemant Kumar[1,2]*, Subodh Srivastava[2] and Balwinder Singh[3]

[1]Department of Physical Sciences, Banasthali Vidyapith, Newai, India
[2]Department of Physics, Vivekananda Global University, Jaipur, India
[3]Department of ACS Division, Centre for Development of Advanced Computing (C-DAC), Mohali, India

Abstract

Memory is remarkably imperative part of any digital system that occupies the most of the part of the design, it is the main issue for designers to design the memory module in such a way that consumes low power and provides stable output so that system performance can be upgraded. In this chapter we proposed the design and implementation of different CNTFET based SRAM memory cell design structures like 6T, 7T, 8T, 9T and 10T at 32nm technology with the help of Cadence Virtuoso tool for ultra-low power applications. This proposed chapter focus on comparative performance design metrics of different CNTFET based SRAM cell structures and proposed design is also compared with existing MOS SRAM cell structures [1], with same design environment.

This design has significant improvement in average power dissipation i.e. 33.4% in 6T, 54.46% in 7T, 40.65% in 8T, 35.74% in 9T & 11.32% in 10T CNTFET SRAM memory cell in comparison of MOS SRAM cells. Performance comparison of CNTFET SRAM memory cells shows that 7T CNTFET SRAM cell is having minimum power dissipation i.e. 7.65nw or 39.85% better than the basic 6T CNTFET based SRAM memory cell and delay, while 9T CNTFET based SRAM memory cell has lowest leakage power dissipation i.e. 75.57pw or 44.64% better than basic 6T CNTFET SRAM cell. Variation in average power and static power dissipation is also measured with respect to parametric variation i.e. variation of chiral vector,

**Corresponding author*: hemantkumar@banasthali.in

Arun Kumar, Manoj Gupta, Mahmoud A. Albreem, Dac-Binh Ha and Mohit Kumar Sharma (eds.)
Wearable and Neuronic Antennas for Medical and Wireless Applications, (137–162) © 2022 Scrivener Publishing LLC

channel length, temperature, and supply voltage, which shows that optimum selection of design parameters can provide fast and power efficient SRAM memory cell through which system efficiency can be enhanced. Cell stability that is also most critical feature of SRAM is also examined for different CNTFET SRAM cell topologies at the supply voltage of 0.7V, keeping pull-ratio (PR) and cell-ratio (CR) of 1. Variation in stability design metrics is also measured with respect to supply voltage in this proposed work. Hence from the result it is noted that proposed design has significant improvement in SRAM memory cell parameters by replacing the existing MOS technology by CNTFET device, utilizing ultra-low power applications.

Keywords: CNTFET, SRAM, static noise margin, stability, N-curve, average power consumption, leakage current

7.1   Introduction

Higher packaging density with good performance and design stability with low power dissipation are the main design constraints for deep submicron technology. As the demands for battery-equipped devices and portable devices is increasing, power dissipation, stability of design has become the critical concerns in digital system design [2]. Power dissipation can be optimized by architecture level optimization, circuit/logic level transform, optimizing circuit speed and device scaling approaches. However, device scaling is the effective and most eminent approach to optimize power dissipation [3], but it has reached its scaling limits, so further scaling down will lead to some serious and challenging issues with respect to fabrication and device performance. Earlier MOS technology was popular; it has approached rapid density growth with scaling of device feature size in recent years [4, 5]. As the device size keeps on downsizing in current MOS technology, critical challenges are faced such as leakage current, short channel effects, reliability issues, and power dissipation. To overcome these issues and limitation of MOS technology, CNTFET is a promising option for planar bulk MOS device [6]. CNTFET device is an ideal candidate which can conquer the restrictions and limitations of MOS based technology with having excellent stability, superior electrical characteristics with high performance and low power dissipation [7].

In modern digital systems, most of the chip area is occupied by the memory part [8]. In SRAM largest portion of the total area and power consumption is covered by basic memory cell [9]. Designing of memory cell must be like that has low power consumption and low read/write time so that all-over performance of a digital system can be enhanced and this can be achieved replacing the existing conventional MOS technology by new emerging technology CNTFET which has the superior current capability, ballistic transport operation and excellent thermal conductivities [10–12]. In this proposed work, different cell structures of CNTFET SRAM are designed then performance and stability is also examined. Comparative analysis of different cell topologies is also performed and results compared with previous work i.e. MOS based SRAM cell design [1] with same design environment. This work shows the significant improvement in SRAM cell design parameters can be achieved by replacing the existing planar MOS SRAM cells by emerging technology i.e. CNTFET. The organization of this chapter as follows, basic operation and structure of different cell topologies is explained in Section 7.2. Then the next section deals with result and comparative performance analysis, Section 7.4 describes about the stability analysis of memory cells. In the last section, a conclusion is stated in Section 7.5.

7.2   CNTFET Based SRAM Memory Cell

Schematic of 6T CNTFET based SRAM memory cell is represented in Figure 7.1 i.e. composed of six transistors, in which N1, N2, P1 and P2 four transistors are connected in cross coupled form and two transistors N4 and N5 transistors called access transistors associated with bit (BL) & bit bar (BLB) lines respectively.

Structure of CNTFET based SRAM memory cell is the alike as that of planar MOS SRAM, just MOS devices are replaced by CNTFET devices. Both access transistors N4 and N5 are activated by a common signal that is WL line, activated for the read & write operation. For the write operation,

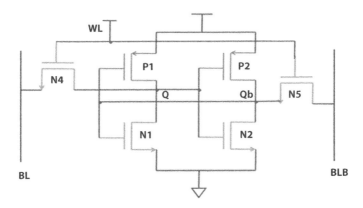

Figure 7.1 Schematic of 6T based CNTFET SRAM memory cell.

bit line BL & bit bar line BLB act as data input line with WL pulled high and at Q & Qb node data will be stored. Data stored at Q & Qb node is accessed with the help of access transistors N4 and N5 in read operation. For read operation the both bit lines BL and BLB are precharged with WL made high. 6T SRAM memory cell is basic memory cell that consumes high power so the comprehensive functioning of the digital system is also affected. Structure of CNTFET based 7T SRAM memory cell is represented in Figure 7.2. An extra transistor N5 provides virtual ground, whenever this will get turn on by pulling high to WL line. However, area will get increased but power dissipation can be optimized by using this structure of memory.

To get notable improvement in design parameters of SRAM memory cell, extra transistors are used with conventional structure of SRAM memory cell. Schematic of 8T_CNTFETSRAM cell is shown as Figure 7.3. It has separate read and write lines are provided namely RWL and WWL to

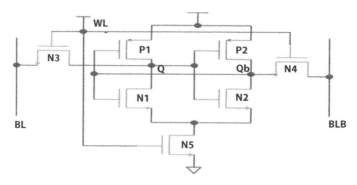

Figure 7.2 Schematic of 7T CNTFET based SRAM memory cell.

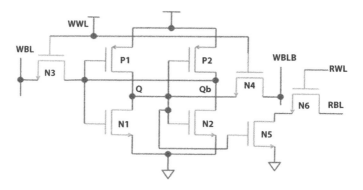

Figure 7.3 Structure of 8T CNTFET based SRAM cell.

perform write/read operation respectively. Before the write and read operation RBL and WBL are pulled high, after that RWL/WWL lines made high to read/write operation respectively. 8T SRAM cell provides more immune to noise voltage and effective memory speed. As we know trade-off exists in power, delay and area, so substantial improvement in power and delay is achieved at the cost of area penalty in comparison of 6T SRAM memory cell.

CNTFET 9T SRAM cell structure is shown in Figure 7.4, extra three transistors are used to get fast and power efficient SRAM with 6T SRAM memory cell.N7 transistor is used to reduce power dissipation by providing virtual ground whenever N7 is 'on'. Structure of this SRAM memory cell is alike to 6T SRAM memory cell except N5, N6, and N7 transistor. N5 and N6 transistors are connected to BL/BLB lines respectively. This 9T SRAM memory cell is used to reduce static power dissipation in standby mode or hold state, that makes SRAM faster and power efficient. Thus, the allover performance of the digital system will get enhanced.

Design of 10T CNTFET based SRAM memory cell structure is shown as Figure 7.5. The memory cell incorporates an additional inverter (N5 & P3 transistor) and a transmission gate (P4 & N6 transistor) along with conventional 6T SRAM memory cell. RE and REB signals are applied at the gate terminal of P4 and N6 transistor as controlling signals, so proper RE/REB signals should be applied to get turn on transmission gate.

The input signal to transmission gate is feed by an inverter, made of P3 and N5 transistors, that is connected by stored node Q of memory cell. For the read operation transmission gate and inverter is used. The biggest advantage of this cell is that there is no necessity to design precharge and sense amplifier circuitry as required in traditional 6T CNTFET based

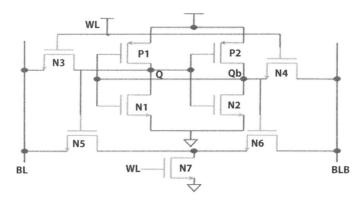

Figure 7.4 Schematic of 9T based CNTFET SRAM memory cell.

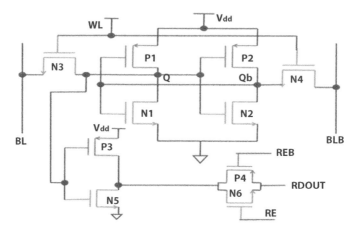

Figure 7.5 Schematic of 10T CNTFET SRAM cell.

SRAM memory cell. Write operation of this CNTFET based memory cell is the same as that of basic SRAM cell. Data applied at bit/bit bar lines will get stored at Q/Qb node with word line made high. Good performance with low power dissipation can be achieved by using additional transistors in comparison of basic 6T SRAM memory cell at the cost of area penalty.

7.3   Simulation Results and Comparative Performance Analysis

This section covers the simulation result of the proposed design. There are different parameters of CNTFET. Here are the list of parameters (Table 7.1) that used for analysis and simulation.

Transient response is investigated as simulation result by adopting common simulation environment using specter simulator at supply voltage at 0.7 V and simulated transient result for 6T, 7T and 9T based CNTFET SRAM memory cell is shown as Figure 7.6. For CNTFET device Verilog-A model, represented by Stanford Compact Model [13], is used to generate symbol of the device with the help of Cadence Virtuoso Tool. Chiral vector that decides the electrical characteristics of device [14] is selected (19, 0). The number of CNT tubes is one, another main important parameter CNT work function 4.5 ev is selected. Transient response as a simulation result for proposed CNTFET based 8T and 10T based SRAM memory cell is shown in Figures 7.7 and 7.8 respectively.

Table 7.1 CNTFET design parameters for this analysis.

Parameter	Description	Values
L_channel	Length of Channel gate	32 nm
Lceff	Average free path for CNT tube	200 nm
L_sd	Doped source/drain region length	32 nm
photon	The optical phonon energy	0.16 eV
L_relax	Carrier relaxation range at drain side, used to match band-to-band tunneling current	40 nm
Leff	Average free path doped CNT	15 nm
Phi__M	Work function of Source/Drain region metal	4.6 eV
Phi__S	Work function	4.5 eV
Pitch	Distance between centers of adjacents CNT	20 nm
(n1, n2)	Chiral vector	(19, 0)
tubes	Number	1

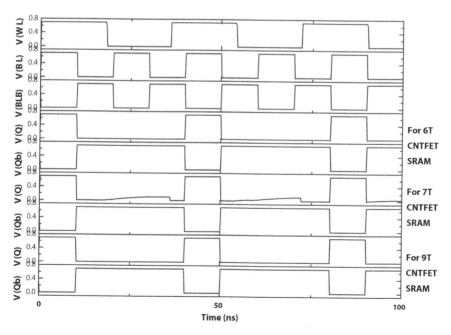

Figure 7.6 Simulation results of 6T, 7T and 9T based CNTFET SRAM memory cell.

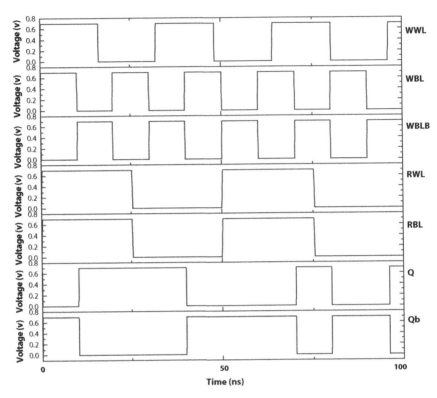

Figure 7.7 Simulated transient response of 8T based CNTFET SRAM cell.

Comparative performance analysis of proposed memory cell topologies and conventional planar MOS based SRAM cells [1] is reported as Table 7.2.

Result of comparative analysis shows CNTFET SRAM cell is fast and power efficient in comparison to conventional SRAM memory cell. Gate leakage current is also observed in the proposed work and also result compared with previous work i.e. conventional MOS SRAM cell [1] that is reported in Table 7.3. This shows that CNTFET SRAM memory cell is having less leakage current in comparison of MOS based SRAM memory cell. Leakage current and leakage power dissipation are observed when access transistors are not working and it is also called hold state of SRAM memory cell in which memory cell will hold the previous data till WL made low [15, 16].

Effect of variation of chiral vector on average power consumption can be shown in Figure 7.9 for 6T, 7T, and 8T SRAM memory cell, which shows the power consumption getting increased with increasing chiral vector. So, the proper value of the chiral vector should be selected to get the optimum value of power dissipation.

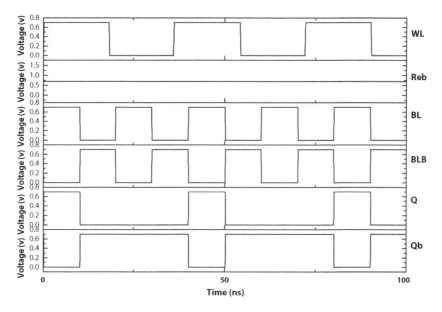

Figure 7.8 Simulation results of 10T CNTFET based SRAM cell.

Average power dissipation variation is also observed with respect to channel length, that can be shown in Figure 7.10. As the length of channel is increasing power dissipation also increasing, so selection of channel length has also critical concern to decide the performance the system.

Variation in average power dissipation is also possible along with variation of power supply voltage that can be shown as Figure 7.11. 6T CNTFET SRAM based memory cell has power consumption, at power supply voltage 700 mv, is 1.27E–08 W or 12.7 nw and when supply voltage is increased then power consumption is also increasing, at supply voltage 5 V power consumption is 1.54E–03 W or 1.54 mw. Power also gets affected with increasing temperature; Figure 7.12 illustrates the effect of variation of average power consumption w.r.t. temperature.

Leakage power also can be varied with the variation of supply voltage, channel length, chiral vector and temperature. Figures 7.13–7.15 show the effect of variation of CNTFET device design parameters like channel length, chiral vector and supply voltage on leakage power for different topologies of proposed SRAM cell. Delay is also critical design parameter that significantly impacts the performance of digital system design. Performance can be enhanced by adjusting design parameters those are affecting to delay. Variation in delay with respect to power supply voltage for proposed design is observed that can be represented in Figure 7.16.

Table 7.2 Comparative analysis of performance for proposed SRAM memory cell structures.

Parameter/Structure		6T SRAM	7T SRAM	8T SRAM	9T SRAM	10T SRAM
Technology	Present work	32 nm	32 nm	32 nm	32 nm	32 nm
	Previous work [1]	45 nm	45 nm	45 nm	45 nm	45 nm
Supply Voltage (v)	Present & Previous Work [1]	700 mv	700 mv	700 mv	700 mv	700 mv
Average Power Dissipation (w)	Present work	12.72 nw	7.65 nw	14.48 nw	11.63 nw	10.73 nw
	Previous work [1]	19.1 nw	16.8 nw	24.4 nw	18.1 nw	12.1 nw
Delay0(s)	Present work	10.07 ns	24.19 ps	34.93 ps	10.12 ns	29.68 ps
Delay1(s)		96.06 ps	89.49 ps	10.14 ps	50 ps	96.45 ps
Previous work delay [1]		110 ps	92.8 ps	81.7 ps	84.4 ps	50 ps
PDP0 (w*s)	Present work	128.26a	185.2z	505.8z	117.7a	318.6z
PDP1(w*s)		1.22a	685.2z	146.8a	581.4z	1.035a
Leakage Power Dissipation (w)		136.53 pw	131.9 pw	178.5 pw	75.57 pw	196.9 pw

Table 7.3 Comparative analysis of leakage current.

Leakage current by transistor		6T SRAM	7T SRAM	8T SRAM	9T SRAM	10T SRAM
N1	Present work	39.18fA	24.52pA	5.65fA	22.9fA	25.69fA
	Previous work [1]	4.103nA	3.32nA	4.7nA	4.18nA	10.555nA
N2	Present work	10.37fA	1.363fA	22.69fA	8.148fA	6.59fA
	Previous work [1]	5.362nA	4.938nA	5.60nA	1.34nA	1.807nA
P1	Present work	30.67aA	4.77fA	47.72fA	6.72fA	105.8aA
	Previous work [1]	4.22nA	2.03nA	2.87nA	6.15nA	671.2pA
P2	Present work	83.12fA	24.54pA	131.9aA	40.96fA	53.16fA
	Previous work [1]	11.83nA	2.51nA	4.742nA	5.509nA	8.656nA

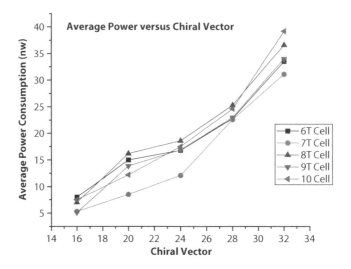

Figure 7.9 Average power consumption versus chiral vector.

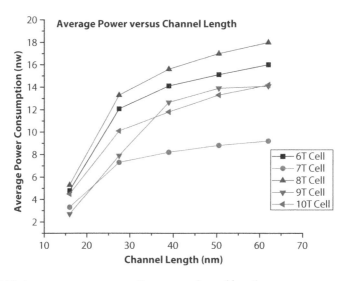

Figure 7.10 Average power consumption versus channel length.

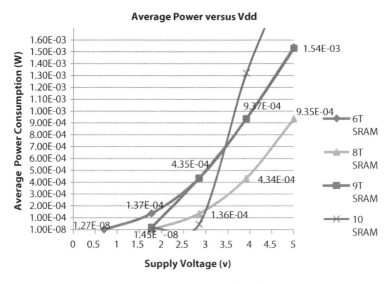

Figure 7.11 Average power consumption versus supply voltage.

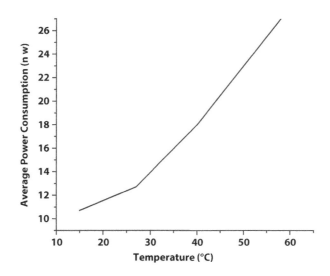

Figure 7.12 Average power consumption versus temperature.

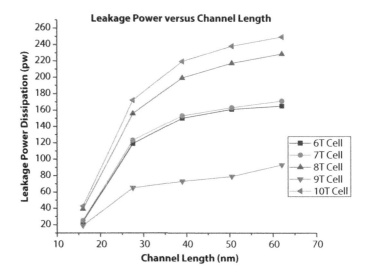

Figure 7.13 Leakage power versus channel length.

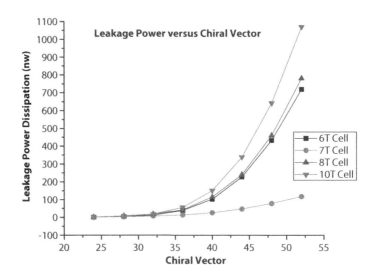

Figure 7.14 Leakage power v/s chiral vector.

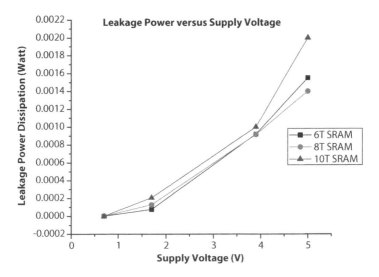

Figure 7.15 Leakage power versus supply voltage.

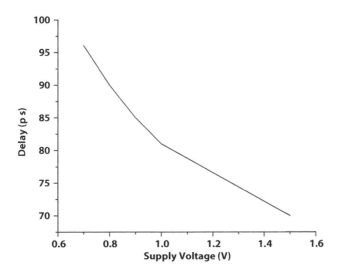

Figure 7.16 Delay versus supply voltage.

7.4   Stability Analysis of Proposed SRAM Cells

SNM is the key figure merit for SRAM memory cell, that decides the stability of the cell. Cell stability, expressed by SNM is an important aspect of SRAM cell design, obtained by maximum possible square in voltage transfer curves (VTC) of CMOS inverters [17] as represented in Figures 7.17–7.19 for hold mode, read mode and write mode. Good SNM of memory cell can be obtained by appropriate power supply voltage, pull up ratio (PR) and cell ratio (CR) [18]. The SNM of SRAM cells increased with increasing the supply voltage [19]. SNM of memory cell indicates about the ability to tolerance of maximum amount of noise [20]. In the proposed memory cell structures stability of cell is investigated by graphical approach i.e. drawing the butterfly curve for SRAM cell in read, write and hold mode to find out read SNM (RSNM), hold SNM (HSNM) and write SNM (WSNM). RSNM, HSNM and WSNM define the ability of SRAM memory cell to read, hold and write the data respectively. Parameter chosen for calculation of cell stability is supply voltage 0.7 V, temperature 27 °C and CR = PR = 1.

Table 7.4 shows measured values of SNM for proposed cell topologies. The measure value indicating that if external noise is larger than the measured value of SNM the SRAM memory cell can lost the data.

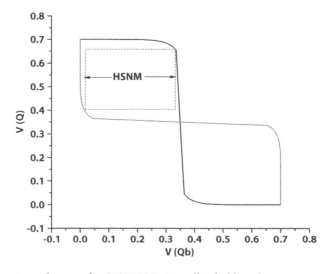

Figure 7.17 Butterfly curve for CNTFET SRAM cell in hold mode.

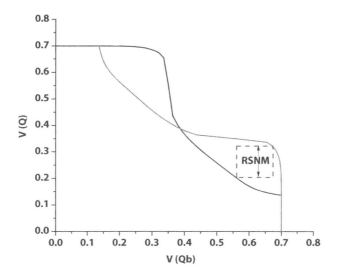

Figure 7.18 Butterfly curve for SRAM in read mode.

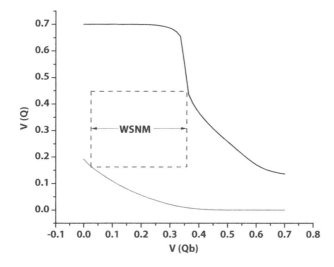

Figure 7.19 Butterfly curve for SRAM in write mode.

Memory cell stability i.e. expressed by SNM variation with supply voltage is also observed in the proposed work, that can be shown in Figures 7.20–7.22.

However, cell stability is measured with the help of graphical approach i.e. by drawing the butterfly curve, but this method doesn't provide

Table 7.4 Static Noise Margin for proposed SRAM cells.

SNM/SRAM cell topology	6T SRAM	7T SRAM	8T SRAM	9T SRAM	10T SRAM
HSNM (mv)	300	285	276	291	286
RSNM (mv)	130	120	110	118	111
WSNM (mv)	340	380	345	342	326.5

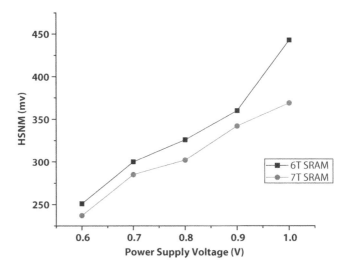

Figure 7.20 HSNM variations with supply voltage.

complete functional information for read and write stability. This method only provides SNM, to measure Static Current Noise Margin (SINM) further more mathematical calculations are required [21–29]. So, another approach to measure the complete information of cell stability of proposed work N-curve method is employed. This N-curve method defines the current and voltage information for the read stability and write stability by four design metrics these are Static Voltage Noise Margin (SVNM), Static Current Noise Margin (SINM), Write Trip Voltage (WTV) and Write Trip Current (WTI).

N-curve for proposed 6T SRAM cell is shown in Figure 7.23, where the voltage difference between P and Q is defined by SVNM and peak value of current between these points is termed as SINM. Similarly, the voltage difference between Q and R is called WTV and negative peak current between these points is WTI. Read stability is characterized by SVNM and

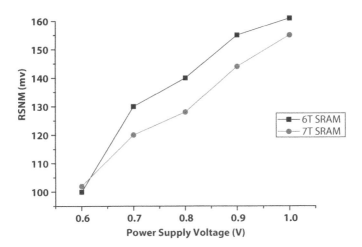

Figure 7.21 RSNM variations with power supply.

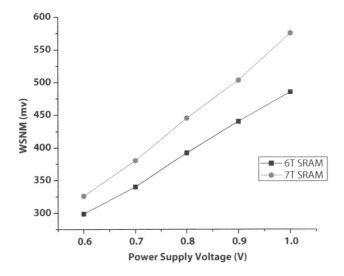

Figure 7.22 WSNM variations with power supply.

SINM while write stability is defined by WTI and WTV. Table 7.5 shows the stability metrics for different CNTFET SRAM cell structures of SRAM cells at supply voltage of 0.7 V and 27 °C temperature. Stability design metrics can be varied with supply voltage, this variation is also observed for the proposed design, are shown in Figures 7.24–7.27.

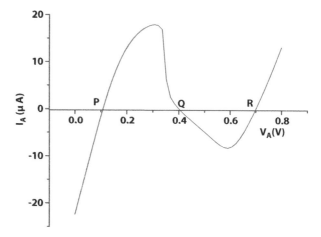

Figure 7.23 N-curve for proposed 6T SRAM cell.

Table 7.5 Stability metrics analysis of proposed SRAM cells using N-curve.

SNM/SRAM cell topology	6T SRAM cell	7T SRAM cell	8T SRAM cell	9T SRAM cell	10T SRAM cell
SVNM (mv)	288	256	317	284	294
SINM (µA)	17.93	12.2	22.5	17.7	18
WTV (mv)	304	288	341	307.5	309
WTI (µA)	-8.24	-5.7	-12.6	-8.0	-8.3

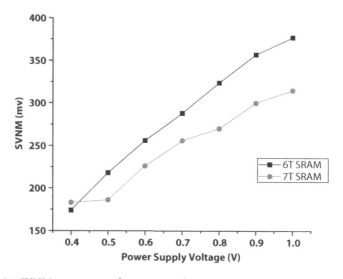

Figure 7.24 SVNM variations with power supply.

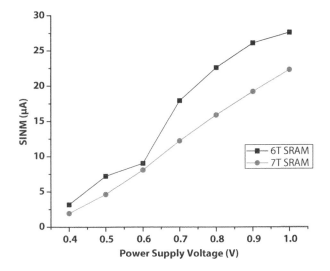

Figure 7.25 SINM variations with power supply.

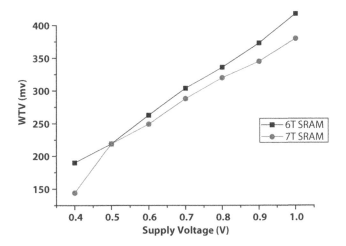

Figure 7.26 WTV variations with supply voltage.

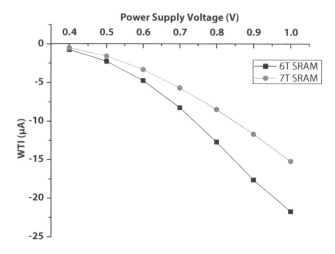

Figure 7.27 WTI variations with supply voltage.

7.5   Conclusion

In this research work different CNTFET based SRAM memory cell topologies are designed then comparative performance and stability analysis is carried out. Result shows that proposed designed structures dissipates the less power and provides high stability with enhanced performance in comparative to conventional planar MOS based SRAM memory cells. Results are also compared with previous work [1] using same design environment, which shows improvement of 33.4% in 6T, 54.46% in 7T, 40.65% in 8T, 35.74% in 9T, 11.32% in 10T SRAM cell, in average power consumption and significant reduction in leakage current is also achieved just by replacing conventional MOS from new emerging technology CNTFET. Comparison of design metrics of different CNTFET SRAM memory cell shows that 7T CNFET SRAM memory cell has the minimum average power consumption i.e. 7.65 nw or 39.85% less than the basic 6T SRAM cell and delay, while 9T CNTFET SRAM memory cell is having minimum leakage power dissipation i.e. 75.57 pw or 44.64% better than basic 6T SRAM cell among them. Effect of variation of temperature, chiral vector, channel length and supply voltage on average power and leakage power consumption are measured for different structures of CNTFETs based SRAM memory cells that motivate the designer for optimum selection of design parameters to enhance the performance of digital system. Stability analysis is also carried out to examine the cell stability by using butterfly curve and N-curve method in this proposed work. 6T SRAM cell has highest HSNM and RSNM i.e. 300 and 130 mv respectively, while 7T SRAM cell has highest WSNM among different structures of SRAM cell designed in this paper. Stability parameters of the SRAM memory cell also get varied with respect to power supply voltage that can affect the stability of memory which is the most important design metrics of SRAM cell. Variation in stability parameters also observed with respect to supply voltage that shows the cell stability can also be varies with respect to supply voltage.

References

1. Vamsi, P.N. and Saxena, N.K., Design and Analysis of Different Types SRAM Cell Topologies, in: *Proceedings of IEEE International Conference on Electronics and Communication System*, p. 1060, 2015, doi: 10.1109/ECS.2015.7124742.

2. Kumar, H., Srivastava, S., Singh, B., Comparative Analysis of 6T, 7T Conventional CMOS and CNTFET based SRAM Memory Cell Design. *J. Adv. Sci. Eng. Med.*, 11, ½, 3, 2019. doi.org/10.1166/asem.2019.2301.

3. Hu, C., Device and Technology Impact on Low Power Electronics, in: *Low Power Design Methodologies*, J.M. Rabey, and M. Pedram, (Eds.), Kluwer Academic Publishers, Norwell, U.S.A., pp. 21–36, 1996.

4. Nishant, P., Deng, J., Mitra, S., Wong, H.S.P., Circuit-level performance benchmarking and scalability analysis of carbon nanotube transistor circuits. *IEEE Trans. Nanotechnol.*, 8, 1, 37, 2009, doi: 10.1109/TNANO.2008.2006903.

5. Morifuji, E., Patil, D., Horowitz, M., Nishi, Y., Power Optimization for SRAM and Its Scaling. *IEEE Trans. Electron Devices*, 54, 4, 715, 2007. doi: 10.1109/TED.2007.891869.

6. Keshavarzi, A., Raychowdhury, A., Kurtin, J., Roy, K., De, V., Carbon nanotube field-effect transistors for high-performance digital circuits—Transient analysis parasitics and scalability. *IEEE Trans. Electron Devices*, 53, 11, 2718, 2006. doi: 10.1109/TED.2006.883813.

7. Pushkarna, A., Raghavan, S., Mahmoodi, H., Comparison of performance parameters of SRAM designs in 16 nm CMOS and CNTFET technologies. *Proceedings of IEEE International SOC Conference*, p. 339, 2010, doi: 10.1109/SOCC.2010.5784690.

8. Kang, S.J., Kocabas, C., Ozel, T., Shim, M., Pimparkar, N., Alam, M.A., Rotkin, S.V., Rogers, J.A., High-performance electronics using dense perfectly aligned arrays of single-walled carbon nanotubes. *Nat. Nanotechnol.*, 2, 230, 2007. doi.org/10.1038/nnano.2007.77.

9. Yeo, K.-S. and Roy, K., *Low-Voltage, Low-Power VLSI Subsystems*, McGraw-Hill Education, India, 2009.

10. Murotiya, S.L., Matta, A., Gupta, A., Performance Evaluation of CNTFET-Based SRAM Cell Design. *IJEEE*, 2, 1, 1, 2012.
11. Singh, A., Khosla, M., Raj, B., Comparative Analysis of Carbon Nanotube Field Effect Transistor and Nanowire Transistor for Low Power Circuit Design. *J. Nanoelectron. Optoelectron.*, 11, 3, 388, 2016. doi.org/10.1166/jno.2016.1913.
12. Singh, A., Khosla, M., Raj, B., CNTFET Modeling and Low Power SRAM Cell Design. *Proceedings of the IEEE 5th Global Conference on Consumer Electronics*, p. 1, 2016, doi: 10.1109/GCCE.2016.7800437.
13. Stanford University Nanoelectronics Group, *Stanford University CNFET Model*, Retrieved from Nov. 21, 2021, https://nano.stanford.edu/stanfordcnfet-model-verilog.
14. Sheng, L., Kim, Y.-B., Lombardi, F., Design of a CNTFET-based SRAM cell by dual-chirality selection. *IEEE Trans. Nanotechnol.*, 9, 1, 30, 2010. doi: 10.1109/TNANO.2009.2025128.
15. Sominenia, R.P., Madhavib, B.K., LalKishorec, K., Low Leakage CNTFET SRAM Cells. *Proceedings of the Elsevier 3rd International Conference on Recent Trends in Computing (ICRTC)*, p. 1049, 2015, https://doi.org/10.1016/j.procs.2015.07.379.
16. Rajendra Prasad, S., Madhavi, B.K., Lal Kishore, K., Low Leakage-Power SRAM Cell design using CNTFETs at 32 nm Technology. *Proceedings of the International Conference on Advances in Communication, Network, and Computing (Springer LNICST)*, vol. 108, p. 165, 2012.
17. Lim, W., Chin, H.C., Lim, C.S., Tan, M.L., Performance evaluation of 14 nm FinFET-based 6T SRAM cell functionality for DC and transient circuit analysis. *J. Nanomater.*, 105, 17, 1–8, 2014. https://doi.org/10.1155/2014/820763.
18. Kureshi, A.K. and Hasan, M., Performance comparison of CNFET- and CMOS-based 6T SRAM cell in deep submicron. *Microelectronics J.*, 40, 6, 979, 2009. https://doi.org/10.1016/j.mejo.2008.11.062.
19. Kumar, S., Tikkiwal, V.A., Gupta, H.O., Read SNM free SRAM cell design in deep submicron technology. *Proceedings of the International Conference on Signal Processing and Communication (ICSC-2013)*, p. 375, 2013, doi: 10.1109/ICSPCom.2013.6719816.
20. Kumar, H., Srivastava, S., Singh, B., Low power, high-performance reversible logic enabled CNTFET SRAM cell with improved stability. *Mater. Today: Proc.*, 42, 4, 1617, 2020. doi.org/10.1016/j.matpr.2020.06.475.
21. Anitha, D., Chari, K.M., Kumar, P.S., N-Curve analysis of low power SRAM Cell. *Second International conference of Inventive Communication and Computation technologies (ICICCT)*, p. 1645, 2018, doi:10.1109/ICICCT.2018.8473215.
22. Gupta, S., Gupta, K., Pandey, N., Stability Analysis of Different Dual-Port SRAM cells in Deep Submicron Region using N-Curve Method. *Proceedings of the IEEE International Conference on Signal Processing and Communication (ICSC)*, p. 431, 2016, doi: 10.1109/ICSPCom.2016.7980619.

23. Gupta, N., Pahuja, H., Singh, B., Nagpal, N., A Novel Design of Low Power Nonvolatile 10T1R SRAM Cell. *Proceedings of the IEEE International Conference on Wireless Netwoks and Embedded Systems*, p. 1, 2016, doi: 10.1109/WECON.2016.7993482.

24. Kulkarni, J.P., Goel, A., Ndai, P., Roy, K., A Read-Disturb- Free Differential Sensing 1R/1 W Port 8T Bitcell Array. *IEEE Trans. Very Large Scale Integr. (VLSI) Syst.*, 19, 9, 1727, 2011. doi: 10.1109/TVLSI.2010.2055169.

25. Spasova, M., Nikolov, D., Angelov, G., Radonov, R., Hristov, M., SRAM design based on carbon nanotube field effect transistor's model with modified parameters. *Proceedings of the IEEE 40th International Spring Seminar on Electronics Technology (ISSE)*, p. 1, 2017, doi: 10.1109/ISSE.2017.8000953.

26. Strangio, S., Palestri, P., Esseni, D., Selmi, L., Crupi, F., Richter, S., Zha, Q.T., Mantl, S., Impact of TFET Unidirectionality and Ambipolarity on the Performance of 6T SRAM Cells. *IEEE J. Electron Device Soc*, 3, 3, 223, 2015. doi: 10.1109/JEDS.2015.2392793.

27. Chek, D.C.Y., Tan, M.L.P., Ahmadi, M.T., Ismail, R., Arora, V.K., Analytical modeling of high performance single walled carbon nanotube field-effect-transistor. *Microelectronics J.*, 41, 9, 579, 2010. https://doi.org/10.1016/j.mejo.2010.05.008.

28. Saun, S. and Kumar, H., Design and Performance Analysis of 6T SRAM Cell on Different CMOS Technologies with Stability Characterization. *IOP (MSE) Proceedings of International Conference on Materials Science and Manufacturing Technology*, vol. 561, p. 1, 2019, doi: 10.1088/1757-899X/561/1/012093.

29. Kumar, H. and Saun, S., Power Gated Technique to improve design metrics of 6T SRAM Memory Cell for Low Power Applications. *ICTACT J. Microelectronics*, 5, 3, 815, 2019.

8

Arrow Shaped Dual-Band Wearable Antenna for ISM Applications

Mehaboob Mujawar

Goa College of Engineering, Ponda-Goa, India

Abstract

In this chapter, we have designed arrow shaped dual-band wearable antenna. This antenna operates in the Industrial, Scientific, Medical [ISM] band of frequencies i.e. 2.4 and 5.8 GHz. The antenna offers low backward radiations at 2.4 and 5.8G Hz, which is considered to be of great concern, as low backward radiations will reduce the EMI effects on the body. Feeding is done using coplanar-waveguide configuration. The reliability of the antenna can be increased by analyzing Specific Absorption Rate [SAR] at the specified dual-band frequencies and safety of the product for human use was ensured. This antenna was designed in CST software. Rogers's substrate was used to build the antenna, which contains LCP material, making it suitable for wearable applications.

Keywords: Microstrip patch, ISM band, dual-band, Roger's substrate, SAR

8.1 Introduction

Wearable antennas have attracted the attention of many researchers, due to its advancement and demand for wearable antennas in market. It has wide range of applications, starting from tracking applications to remote health care sector. The research made in the field of wearable antennas has resulted in the product being compact, flexible and which can operate on multi-band frequencies. Antenna design with Meta-materials has also gained importance by improving the performance of the antenna.

Email: mehaboob311134@gmail.com

Arun Kumar, Manoj Gupta, Mahmoud A. Albreem, Dac-Binh Ha and Mohit Kumar Sharma (eds.) *Wearable and Neuronic Antennas for Medical and Wireless Applications*, (163–178) © 2022 Scrivener Publishing LLC

The introduction of a slot on the patch makes it suitable to work on multiband frequencies. Various feeding techniques can be used to improve the performance of the antenna. Wearable antennas are basically smart wearables which have a wide range of applications in the scientific, medical and industrial sector. We can consider some examples of smart wearables which we use in our everyday life such as Smart watch, Google glass, and wearable Cameras. Smart watch comprises of Bluetooth antennas which are blended on its surface. Google glass makes use of special Wi-Fi and GPS antennas. Wearable cameras make use of Bluetooth and Wi-Fi antennas to capture images. Antennas can also be embedded in shoes, which communicate to smartphone via Bluetooth. When we deal with wearable devices the concept of Wireless Body Area Network [WBAN] plays a very important role. Wireless Body Area Network has been defined with an international standard IEEE 802.15.6. WBAN contributes to the implementation of wearable devices in the field of consumer and medical electronics. IEEE 802.15.6 standard supports wide range of data rates for ISM applications. This standard aims at providing secure environment for the user by restricting specific values for power and range of the wearable devices. There are three types of wearable devices. On-body devices are usually body-worn devices, which the user can use as per his convenience. Off-body devices are those, which are placed in the external wearable. Antenna implanted in shoes, clothing can be considered as the example of Off-body devices. In-body devices are those, which are implanted in the human body.

There are many challenges, which are associated with wearable devices. The first challenge associated with wearable device is the location, where the antenna has to be placed on the human body. When the wearable device starts to radiate, it emits electromagnetic radiations. The efficiency will decrease due to the placement of the antenna on human body. For example, if we consider the case of an antenna which initially gives efficiency of 50%. But the same antenna placed on human body will give efficiency of 5%. This indicates that wearable devices have to be designed properly, since it shows large variations in its working depending on the placement of antenna on the human body. The second challenge of wearable device is its compactness. The main aim of the antenna designers is to reduce the size of antenna on the wearable device. While designing asmart watch, we cannot randomly consider big dipole antenna to implant on it. Hence the selection of the type of antenna and its dimensions are of very much importance during the design stage. The performance of the antenna has to be stable under different atmospheric conditions. It is also desirable to have wearable devices which operate on multiple frequency bands, hence enabling the operation of the devices under different applications. It is of great concern for the design engineers to reduce the dimensions of the wearable device keeping constant all its operational features.

The antenna design considerations have to be taken into account in order to enhance the working of wearable antenna. It is desirable to have antenna with low profile. It is preferable to use antennas orthogonal to the body than the parallel antennas. When wearable antenna is implanted on the human body, it results in the detuning of the antenna due to loading effect. Therefore, the designer needs to supervise the changes occurring in the resonating frequency band because of this effect. Ground plane creates a shield around the human body to protect it from radiations. These wearable antennas need to be portable and compact, since they will be used for human body wearable applications. These devices also need to be economical, so that they can be used by all sections of society. These antennas should also be flexible and comfortable to use on any part of the human body. The antenna parameters such as reflection coefficient, resonant frequency and bandwidth will be affected by crumpling of the wearable device. Crumpling creates detuning effect on the antenna. These antennas need to have higher radiation efficiency, when implanted on human body, to reduce the losses. Patch antenna is mainly used because of its low profile, higher directivity and many other advantages. These can be easily fabricated on the printed circuit boards and it tends to be one of the reasons for its large usage in the construction of portable devices.

In literature, various types of antennas operating in dual-band have been studied. Paper [1] presents conventional design of antennas that is by making use of printed circuit boards, which is not acceptable by few devices because of various antenna performance parameters. For efficient working of antenna and to protect the tissues from damage, due to on body devices, antennas should be implanted on the outer layers of clothing. It is also comfortable for the user, when wearable antennas are placed on the textiles rather than body implant. Paper [2] presents flexible textile antenna, which proves to be more effective compared to traditional metallic antennas. Planar Inverted F antenna has been used to obtain wider antenna band operations. A wearable PIFA was designed using conductive plate and was placed on FlexPIFA substrate at 2.45 GHz. Paper [3] presents meta-material based Left Handed structures, these structures along with LH modes help in the miniaturization of the antenna. This antenna is operating on dual frequencies. This antenna mainly consisted of right T-handed resonator and left T-handed resonator. Paper [4] presents the multi-band antennas, these antennas are widely used in wearable devices depending on the specific application. Flexible antennas also find a wide range of application in RFID and sensing. Paper [5] presents multi-frequency antennas, recent development of multi-frequency antennas have increased their demand in market. These antennas are designed to operate on various operation modes over different wireless services. In order to increase the performance of this antenna, it will be effective to have two radiation patterns mainly dipolar and monopolar [5]. Design presented in paper [5] was compact and low profile. Paper [6] presents dual-band textile antennas; it is the most preferred antenna for wireless applications. However, this antenna has a larger surface area, which sets a limitation for its use directly on body. Specifically on chest or back, but still this antenna has proven to be effective in performance. The fabrication complexity of this antenna has been increased due to the presence of vias on the designed antenna. Paper [7] presents wearable device used for 5G, it is simulated at 28 and 38 GHz. This planar inverted F antenna has been placed on the substrate. The substrate used in this antenna design is jeans. Detailed analysis has been done for measuring the SAR values for both the antenna and smart watch. Paper [8] presents Meta material based dual band wearable antenna, this antenna operates at 2.4 and 5.2 GHz. Jeans is used as substrate. The performance is analyzed in all possible orientations. Paper [9] presents the design of textile antenna, which mainly focuses on the Bluetooth application. This textile antenna was designed in order to provide better performance without affecting the textile properties. This textile antenna has been provided with microstrip line feed. Paper [10] presents a planar antenna, which has been mainly used in telemetry applications. The feeding technique used in this planar antenna is coaxial cable. Paper [11] presents the UWB antenna design that mainly focuses on the antenna design by employing the LCP substrate. LCP substrate helps to improve the antenna performance by allowing the antenna to operate in the frequency range of 2.9 to 10.2 GHz. Paper [12] presents the effects and measurement of radiations from the wearable antennas. The human body tissues absorb electromagnetic radiations, when wearable antenna radiates. Therefore, it is necessary to protect the human body from exposure to high radiations from the antenna. As a solution to this high radiation exposure issue, it is necessary to measure SAR [Specific Absorption Rate] of all the wearable devices before manufacturing in market. SAR values have been restricted by regulatory authorities within the region. According to Federal Communications Commission in US, SAR value is 1.6 W/kg for 1 g of tissue and in Europe, SAR value is 2.0 W/kg for 10 g of tissue [13]. With the advancement and research in the field of wireless communications, the demand for antenna design in millimeter wave applications has also been increasing. Paper [14] presents the additive manufacturing techniques for antenna designs; this technique has a wide number of advantages such as materials required for the fabrication process are less expensive, less wastage of materials. Paper [15, 16] presents UWB antenna for wearable 5G applications, many researchers have contributed in the development of wearable devices for 5G applications and the main point of attraction for antenna engineers is use of flexible antennas in wearable applications. Flexible antennas can be easily implanted on human body [17–20].

In this chapter, Arrow shaped dual band antenna has been designed. This antenna has been specifically designed to operate in the Industrial, Scientific, Medical band of frequencies. The proposed antenna is operating in dual-bands i.e.2.4 and 5.8 GHz, which is suitable for wireless and wearable applications. Various techniques can be used for making dual-band antenna namely use of fractal structures, slots and slits, shorting pins and so on. We have introduced slots on the antenna, so that it could operate in ISM band of frequencies, which is suitable for wearable devices. In literature survey, we had seen many techniques for the implementation of the antennawhich made use of textile material as substrate. But for the proposed antenna design we have used Rogers Ultralam 3850 as substrate, due to some limitations of the textile substrate in our proposed design. In the proposed design, there is a need of antenna, which is flexible, robust and high resistance to outside effects like humidity. These come into effect with the introduction of Roger's substrate.

8.2   Antenna Design

The antenna design is a multi-stage process, which includes initially mathematical calculations to be carried out, in order to determine the dimensions of the antenna according to the required specifications and then finally obtaining the simulation results using software's. The dimensions of the antenna can be calculated by using the following equations.

$$W = \frac{c}{2f}\sqrt{\frac{2}{\varepsilon_r + 1}} \tag{8.1}$$

$$L = L_{eff} - 2\Delta L \tag{8.2}$$

Where W is the width of the patch, c is the velocity of light, f is the resonating frequency, L is the length of the patch, ε_r is the dielectric constant of substrate, L_{eff}, is the effective length of patch and is given by

$$L_{eff} = \frac{c}{2f\sqrt{\varepsilon_{reff}}} \tag{8.3}$$

The normalized extension in length is given by

$$\Delta L = 0.412h \frac{\left(\varepsilon_{reff} + 0.3\right)\left(\dfrac{w}{h} + 0.264\right)}{\left(\varepsilon_{reff} - 0.258\right)\left(\dfrac{w}{h} + 0.8\right)} \tag{8.4}$$

Where ε_{reff} is the effective dielectric constant and is given by

$$\varepsilon_{reff} = \frac{\varepsilon_r + 1}{2} + \frac{\varepsilon_r - 1}{2}\left[1 + 12\frac{h}{w}\right]^{-\frac{1}{2}} \tag{8.5}$$

The width and length of the Substrate can be determined using the formulas given below

$$L_g = L + 6h \tag{8.6}$$

$$W_g = W + 6h \tag{8.7}$$

W_g is the width and L_g is the length of substrate and h is given by

$$h = \frac{0.0606\lambda}{\sqrt{\varepsilon_r}} \tag{8.8}$$

The below mentioned equation can be used to determine the length of the feed line

$$Feed\ length\ \left(L_f\right) = \frac{\lambda_g}{4} \tag{8.9}$$

Where λ_g is the guided wavelength and is given by

$$\lambda_g = \frac{\lambda}{\sqrt{\varepsilon_{reff}}}$$

(8.10)

Efficiency of the antenna is calculated by

$$\eta = \frac{Gain}{Directivity} \times 100\%$$

(8.11)

Where η is the efficiency of the antenna. The equations for the radiation box are mentioned below

$$Axis\ position = \frac{-\lambda_g}{6} + \frac{-\lambda_g}{6} + \frac{-\lambda_g}{6}$$

(8.12)

$$Length = \frac{\lambda_g}{6} + \frac{\lambda_g}{6} + L_g$$

(8.13)

$$Width = \frac{\lambda_g}{6} + \frac{\lambda_g}{6} + W_g$$

(8.14)

$$Height = \frac{\lambda_g}{6} + \frac{\lambda_g}{6} + h$$

(8.15)

The dimensions of the antenna are calculated by using the above mentioned equations and can be used to design antenna in software to obtain simulation results.

Arrow flexible wearable antenna is shown in Figure 8.1, which is used for ISM applications. This arrow shaped antenna has been connected with two resonators and the antenna is feed using CPW technique. The proposed antenna has two narrow slots, which have helped to reduce the resonant frequency from 2.7 to 2.4 GHz, which is the required frequency for ISM application. It also has two L slots, which are inverted and these slots help to operate at resonant frequency of 5.8 GHz. The antenna is excited to 50 Ω, by optimizing the feedline width and gap connecting feedline and ground. Roger's has been used as a substrate. The total area of the proposed antenna is 31 × 19mm², which is typically small and perfect size for wearable applications.

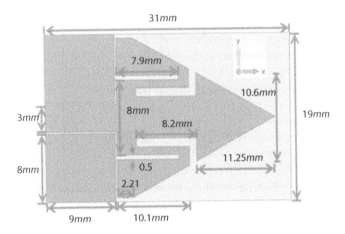

Figure 8.1 Proposed antenna structure.

8.3   Results

The reflection coefficient is a very important parameter of antenna. It helps to determine exactly, how much power has been reflected back from the antenna. Figure 8.2 shows the graph of reflection coefficient v/s frequency for the proposed antenna design. The green dotted line in Figure 8.2 represents the reflection coefficient of the antenna, when the antenna

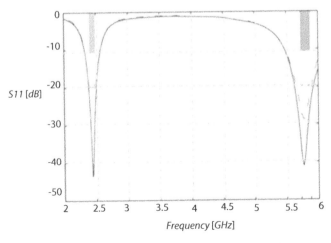

Figure 8.2 On-body [green dotted line] and free space [blue line] reflection coefficient of the proposed antenna.

is placed on human body and blue line represents reflection coefficient of the antenna, when the antenna is placed in free space. As can be observed from Figure 8.2, when the antenna is in free space, the reflection coefficient is −43 dB at 2.4 GHz and −41 dB at 5.8 GHz. Similarly when the antenna is placed on human body, the reflection coefficient is −43 dB at 2.4 GHz and −29 dB at 5.8 GHz.

The reflection coefficient of the antenna during bent condition can be observed in Figures 8.3 and 8.4. When the antenna is bent, it affects working of the antenna to a great extent. There is a need to study the bending

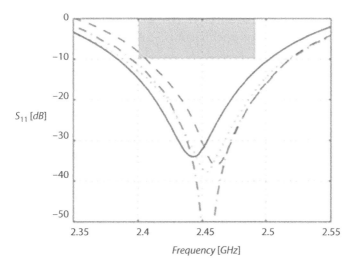

Figure 8.3 Reflection coefficient during bending at 2.4 GHz ISM band.

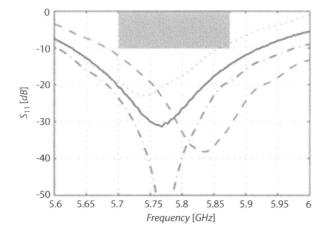

Figure 8.4 Reflection coefficient during bending at 5.8 GHz ISM band.

effects in antenna, so that proper optimization can be carried out for the antenna operating on wide range of frequencies. When the wearable antenna is bent, it decreases the bandwidth of the antenna and also starts resonating at lower frequencies. This effect will be overcomed with the introduction of antennas having wide frequency bandwidths. The bending of the antenna at its width affects the resonant frequency along with electric field and current of the antenna. The working of the antenna will be widely affected when the bending is done at its length. When the antenna is operating at a frequency of 2.4 GHz, free space reflection coefficient is −34 dB and on-body reflection coefficient is −36 dB. When the antenna is operating at a frequency of 5.8 GHz, free space reflection coefficient is −31 dB and on-body reflection coefficient is −38 dB.

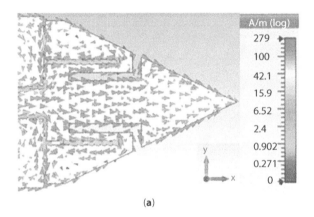

Figure 8.5 (a) Surface current distributions at 2.4 GHz.

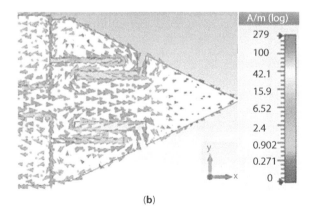

Figure 8.5 (b) Surface current distributions at 5.8 GHz.

Figure 8.5 shows the surface current distributions of the proposed arrow shaped dual-band wearable antenna. Figure 8.5(a) presents the surface current distribution at 2.4 GHz, as can be observed current distribution was more at the arrow shape and at the two narrow slots. Figure 8.5(b) presents the surface current distribution at 5.8 GHz, as can be observed the current distribution was mainly around inverted L slots. The proposed antenna had a gain of 1.12 and 3.12 dB. At resonant frequency of 2.4 GHz, radiation efficiency of the antenna was and at resonant frequency of 5.8 GHz, and radiation efficiency of the antenna was 98%.

8.4   Analysis of Specific Absorption Rate (SAR)

Specific Absorption Rate is defined as the amount of power, which has been absorbed by human tissue per unit mass. SAR values of all the devices must be checked before fabrication process and also before manufacturing the product in market. There is a need to study the effect of human body interaction in wearable antennas. In wearable devices, the resonant frequency of the antenna changes with the change in the permittivity of the body tissues. The permittivity of the body tissues is high, which ultimately leads to detuning of the resonant frequency to a lower value. As the frequency of the wearable device increases, there will be a major effect on the conductivity and relative permittivity of the skin. Therefore, there will be reduction in the relative permittivity and increment in the conductivity of the skin as frequency increases. When the antenna radiates, few radiations will be absorbed by the human body due to its lossy nature and the gain of the antenna will reduce. SAR values have been restricted by regulatory authorities based on the region of application. According to Federal Communications Commission in US, SAR value is 1.6 W/kg for 1 g of tissue and in Europe, SAR value is 2.0 W/kg for 10 g of tissue.

Table 8.1 Comparison of SARs on different part of human body.

Body part	SAR at 2.4 GHz [w/kg]	SAR at 5.8 GHz [w/kg]
Chest	0.012	0.25
Thigh	0.0066	0.21
Upper arm	0.012	0.15
Forearm	0.37	0.69

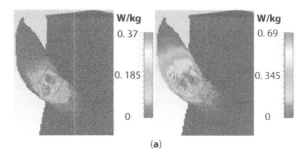

(a)

Figure 8.6 (a) Specific absorption rate of antenna placed on forearm at frequency of 2.4 GHz (left) and 5.8 GHz (right) respectively.

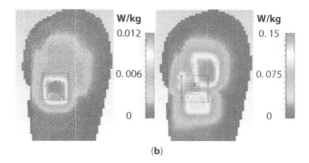

(b)

Figure 8.6 (b) Specific absorption rate of antenna placed on upper arm at frequency of 2.4 GHz (left) and 5.8 GHz (right) respectively.

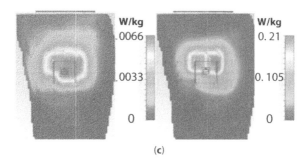

(c)

Figure 8.6 (c) Specific absorption rate of antenna placed on thigh at frequency of 2.4 GHz (left) and 5.8 GHz (right) respectively.

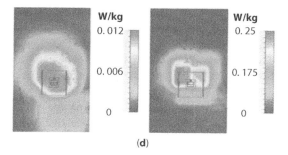

(d)

Figure 8.6 (d) Specific absorption rate of antenna placed on chest at frequency of 2.4 GHz (left) and 5.8 GHz (right) respectively.

The SAR values of the wearable device have been determined by using CST microwave studio while determining the SAR values of the wearable device by locating it on various parts of the human body, the antenna is powered with 0.5 W. A specific standard IEEE C95.1 has been followed for calculating the SAR values. As can be seen in Table 8.1 and also from the simulation results shown in Figure 8.6, we conclude that SAR values were higher at a frequency of 5.8 GHz compared to the SAR Values at 2.4 GHz.

8.5   Conclusion

In this chapter, we have designed compact, dual-band, flexible antenna for wearable application. The overall design of the antenna was of arrow shape. Slots were created on the antenna, so that it could resonate effectively at 2.4 and 5.8 GHz. The observed SAR values indicate that this antenna can be utilized for human wearable applications. The observed antenna parameters such as Gain, radiation efficiency, reflection coefficient have proved to be effective for wearable application.

References

1. Yan, S., Soh, P.J. and Vandenbosch, G.A.E., Made to be Worn. *Electronics Letters*, vol. 50, no. 6, pp. 420, March 2014.
2. Salonen, P., Sydanheimo, L., Keskilammi, M. and Kivikoski, M., A small planar inverted-F antenna for wearable applications. In *3rd Int. Symp. on Wearable Computers Digest*, pp. 95–100, 1999.
3. Caloz, C. and Itoh, T., *Electromagnetic Metamaterials: Transmission Line Theory and Microwave Applications*, New York: Wiley, 2006.

4. Leng, T., *et al.*, Graphene nanoflakes printed felixble meandered-line dipole antenna on paper substrate for low-cost RFID and sensing applications. *IEEE Ant. and Wireless Propagatio Letters*, vol. 15, pp. 1565–1568, 2016.

5. Ahzad, J., Special issue on multifunction antennas and antenna systems. *IEEE Trans. Antennas Propag.*, vol. 54, no. 2, pp. 314–316, Feb. 2006.

6. Yan, S., Soh, P.J. and Vandenbosch, G.A.E., Wearable dual-band composite right/left-handed waveguide textile antenna for WLAN applications. *Electronics Letters*, vol. 50, no. 6, pp. 424–426, March 2014.

7. Jajere, A.M., Millimeter Wave Patch Antenna Design Antenna for Future 5G Applications. *International Journal of Engineering Research & Technology (IJERT)*, vol. 6, no. 2, pp. 298–291, 2017.

8. Agneessens, S. and Rogier, H., Compact Half Diamond Dual-Band Textile HMSIW On-Body Antenna. *IEEE Transactions on Antennas and Propagation*, vol. 62, no.5, pp. 2374–2381, May 2014.

9. Locher, I., Klemm, M., Kirstein, T., Troster, G., Design and Characterization of Purely Textile Patch Antennas. *IEEE Transactions on Advanced Packaging*, vol. 29, no. 4, pp. 777–788, Nov. 2016.

10. Chi, Y.-J., Chen, F.-C., On-Body Adhesive-Bandage-Like Antenna for Wireless Medical Telemetry Service. *IEEE Transactions on Antennas and Propagation*, vol. 62, no. 5, pp. 2472–2480, May 2020.

11. Ur-Rehman, M., Abbasi, Q.H., Akram, M., Parini, C., Design of Band-notched Ultra Wideband Antenna for Indoor and Wearable Wireless Communications. *IET Microwaves, Antennas & Propagation*, vol. 9, no. 3, pp. 243–251, 2015.

12. Hirata, A., Fujiwara, O., Nagaoka, T. and Watanabe, S., Estimation of Whole-Body Average SAR in Human Models Due to Plane-Wave Exposure at Resonance Frequency. *IEEE Transactions on Electromagnetic Compatibility*, vol. 59, pp. 41–48, 2010.

13. Zhao, K., Zhang, S., Chiu, C., Ying, Z. and He, S., SAR Study for Smart Watch Applications. *Antennas and Propagation Society International Symposium (APSURSI)*, Memphis, U.S, 2014.

14. Kumar, A., Albreem, M.A., Gupta, M., Alsharif, M.H., Kim, S., Future 5G Network Based Smart Hospitals: Hybrid Detection Technique for Latency Improvement. *IEEE Access* 8, 153240–153249, 2020

15. Kumar, A., Gupta, M., Le, D.N., Aly, A.A., PTS-PAPR Reduction Technique for 5G Advanced Waveforms Using BFO Algorithm. *Intelligent Automation and Soft Computing* 27 (3), 713–722, 2021

16. Meena, K, Gupta, M., Kumar, A., Analysis of UWB Indoor and Outdoor Channel Propagation. *2020 IEEE International Women in Engineering (WIE) Conference on Electrical and Computer Engineering (WIECON-ECE)*, 352–355, IEEE, 2020.

17. Gupta, M., Chand, L., Pareek, M., Power preservation in OFDM using selected mapping (SLM). *Journal of Statistics and Management Systems* 22 (4), 763–771, 2019.

18. Jun, S., Sanz-Izquierdo, B. and Summerfield, M., UWB antenna on 3D printed flexible substrate and foot phantom. In *Antennas & Propagation Conference (LAPC)*, 2015 Loughborough, pp. 1–5, 2015.
19. Sabban, A., Comprehensive Study of Printed Antennas on Human Body for Medical Applications. *International Journal of Advance in Medical Science (AMS)*, Vol. 1, pp. 1–10, February 2013.
20. Huan, Y. and Boyle, K., *Antenna From Theory to Practice*, John Wiley & Sons, Hoboken, 2008.

.

9

Edge-Fed Semicircular Antenna Enabled With Pins and Slots for Wireless Applications

Mohd Gulman Siddiqui[1]* and Anurag Mishra[2]

[1]Department of Electronics, Banasthali Vidyapith, Newai, India
[2]Department of Physics, Ishwar Saran P.G. College, University of Allahabad, Prayagraj, India

Abstract

This paper proposed an antenna design for the multiband operation using a pair of shorting pins inside the antenna structure using Bakelite substrate. In addition, slot cuts on the patch surface create multiband operation for S-band and X-band applications. The proposed design with shorting pin loaded inside the semicircular patch of the designed antenna works at 2.42, 3.26, 4.55 and 10.43 GHz having percentage bandwidths of 21.59, 13.68, 4.42 and 11.14% respectively. The patch antenna is first simulated using HFSS and then fabricated with LPKF pro tool and also measured and analyzed on the basis of circuit theory conceptand measured results are in agreement.

Keywords: Patch antenna, notch, slot and multiband antenna

9.1 Introduction

Microstrip antenna in recent years has extensively been used in broadband applications like telemetry navigation, satellite communications, mobile technology, biomedical systems, direct broadcast systems (DBS), mainly due to the compactness size, ease of fabrication and low cost of manufacturing. But the patch antenna suffers problems like narrow bandwidth

*Corresponding author: mohdgulmansiddiqui@bansthali.in

Arun Kumar, Manoj Gupta, Mahmoud A. Albreem, Dac-Binh Ha and Mohit Kumar Sharma (eds.) *Wearable and Neuronic Antennas for Medical and Wireless Applications*, (179–190) © 2022 Scrivener Publishing LLC

as well as low. Numerous techniques have been proposed to obtain multiband, antenna loaded with slot, notch, parasitic elements and loading diodes in the patch.

During several years there occurs fast development of multiband antenna for wide application in wireless communication systems. Researchers have reported applications such as Compact shorted microstrip patch antenna [1], a dual band circular patch antenna [2], Half U-slot enabled semicircular patch antenna [3, 4], L-shaped circular disk patch antenna [5], W and inclined slot enabled patch antenna [6, 7]. Also, the multiband antenna is designed for mobile and wireless applications by the same author [18, 19] using high frequency structure simulator HFSS [17].

Here we have proposed a compact pin shorted with slot enabled semicircular patch antenna which exhibits multiband behavior. The designed antenna corporate the return loss for S/X-band behavior. Resonance frequency can bevaried by changing the slot dimensions and shorted pins position.

9.2   Configuration of Proposed Antenna

The proposed geometry of antenna having slots and pin shorted is shown in Figure 9.1. The material used is Bakelite with dielectric constant of 4.78. The antenna is fed using Edge-mount co-axial cable having 50 Ω impedance. In Figure 9.2, the simulation structure for patch using edge mount coaxial connector feeding, both with and without shorting pins.

Analysis of proposed the antenna can be done in similar way like that of the circular patch. Figure 9.3 shows the hardware fabrication of proposed antenna using LPKF pro tool. The resonance frequency [8] is given as

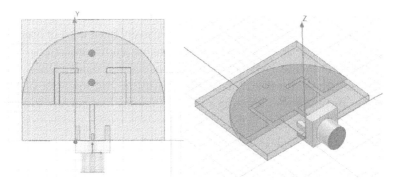

Figure 9.1 The proposed antenna designed in HFSS simulation software.

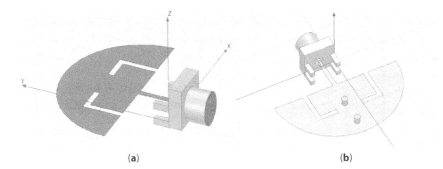

(a)   (b)

Figure 9.2 Edge mount coaxial connector for feeding, (a) without shorting pins and (b) with shorting pins.

Figure 9.3 Fabrication of antenna using LPKF pro tool.

$$f_r = \frac{k_{nm}c}{2\pi a_e \sqrt{\varepsilon_e}} \tag{9.1}$$

Where k_{nm} is the mth zero root of Bessel function derivative having order n and ε_e is the effective dielectric constant of material [9], a_e effective radius given as:

$$a_e = \sqrt{\frac{L_e W_e}{\pi}} \tag{9.2}$$

The effective radius a_e of half of the disk is equated and the area of the disk that needs to be expanded with rectangular patch dimension ($L_e \times W_e$), here L_e and W_e are effective length and effective width of patch and it can be calculated by [10]. The equivalent circuit is shown in Figure 9.4 that calculated the resistance, inductance and capacitance [9].

$$C_1 = \frac{\varepsilon_e \varepsilon_o LW}{2h} \cos^{-2}(\pi x_o / L) \tag{9.3}$$

$$L_1 = \frac{1}{\omega^2 C_1} \tag{9.4}$$

$$R_1 = \frac{Qr}{\omega C_1} \tag{9.5}$$

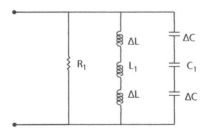

Figure 9.4 The equivalent circuit approach of antenna (notch-enabled).

in which, L is the length of the patch, W is the width of patch, h is the thickness of material, x_0 is feed location, $Q_r = \dfrac{c\sqrt{\varepsilon_e}}{fh}$ and ε_e is the effective permittivity,

$$\varepsilon_e = \frac{\varepsilon_r + 1}{2} + \frac{\varepsilon_r - 1}{2}\left(1 + \frac{10h}{W}\right)^{-1/2}$$

where, ε_r is the relative permittivity.

9.2.1 Analysis of Notch Loading Antenna

Discontinuity due to notch enabled in the patch are considered as a series inductance i.e. ΔL with series capacitance i.e. ΔC which changes the equivalent circuit as in Figure 9.5, and thus the series inductance (ΔL) as well as the series capacitance (ΔC) is calculated [11, 12].

$$\Delta L = \frac{h\mu_0\pi}{8} \tag{9.6}$$

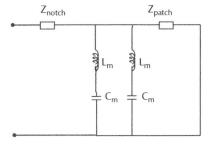

Figure 9.5 The equivalent circuit of patch antenna using notch.

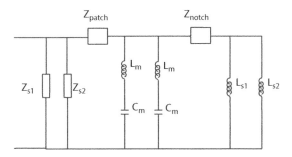

Figure 9.6 the equivalent circuit of shorting pin enabled antenna withslot.

$$\Delta C = \left(\frac{L_n}{L}\right).C_g \tag{9.7}$$

where $\mu_0 = 4 \pi \times 10^{-7}$ H/m
   L_n = depth of the notch
   L = Length of the patch
   C_g = gap capacitance and is given by [13].

The value of resistance R_1, after cutting the notch, is calculated [14]. It is seen that the resonant circuits, after cutting notch is calculated through mutual inductance and capacitance e i.e. L_m and C_m respectively in Figure 9.6.

9.2.2   Analysis of Slots in Antenna

The slots are enabled in the patch, with area $(L_S \times W_S)$, it is be analyzed by duality relationship between the dipole and slots [15]. The radiation resistance can be given as:

$$R_r = \frac{\eta_0 \cos^2 \alpha}{2\pi} \int_0^\pi \left[\frac{\cos\dfrac{k^2 \cos\theta}{2} - \cos\dfrac{kL_S}{2}}{\sin\theta} \right]^2 d\theta \tag{9.8}$$

which yields

$$R_r = 60 \left\{ \begin{array}{l} \left[C + \ln(kL_S)\, C_i(kL_S) + \dfrac{1}{2}\sin(kL_S)\left[S_i(2kL_S)\, 2S_i(kL_S) \right] + \dfrac{1}{2}\cos(kL_S) \right] \\ \left[C + \ln\left(\dfrac{kL_S}{2}\right) + C_i\left(2kL_S\right) 2C_i\left(kL_S\right) \right] \end{array} \right\}$$

Here C is Euler's constant, S_i is sine integral and C_i is cosine integral. Hence, the total input impedance [16] is calculated as:

$$Z_{slot} = \frac{\eta_0^2}{4Z_{cy}} \tag{9.9}$$

in which $\eta_0 = 120\pi\,\Omega$

$$\text{and } Z_{cy} = R_r(kL_S) - j\left[120\left(\ln\left(\frac{L_S}{W_S}\right) - 1\right)\cot\left(\frac{kL_S}{2}\right) - X_r(kL_S)\right]$$

in which L_S and W_S are length and width of the slot respectively.

Here, R_r = real part and it is equal to radiation resistance and X_r = input reactance of the slot [13]. The equivalent circuit, as shown as Figure 9.5. Now, net input impedance of antenna is:

$$Z_L = Z + \frac{Z_m Z_{Patch}}{Z_{Patch} + Z_m} \tag{9.10}$$

Where,

$$Z = \frac{Z_{Slot} Z_{notch}}{Z_{Slot} + Z_{notch}}$$

in where Z_P = input impedance of the patch antenna, and can be calculated:

$$Z_{Patch} = \frac{1}{\dfrac{1}{R_P} + j\omega C_P + \dfrac{1}{j\omega L_P}}$$

$$\text{and } Z_{Slot} = \frac{Z_{S1} Z_{S2}}{Z_{S1} + Z_{S2}}$$

$$\text{and } Z_{notch} = \frac{j\omega R_1 L_2}{j\omega L_2 + R_1 - R_1 L_2 C_2 \omega^2}$$

$$\text{in which } L_2 = L_1 + 2\Delta L, \; C_2 = \frac{C_1 \Delta C}{2C_1 + \Delta C},$$

$$\text{and } Z_m = \left(j\omega L_m + \frac{1}{j\omega C_m}\right)$$

$$L_m = \frac{C_p^2(L_1 + L_2) + \sqrt{C_p^2(L_1 + L_2)^2 + 4C_p^2(1 - C_p^2)L_1L_2}}{2(1 - C_p^2)} \tag{9.11}$$

$$C_m = -\frac{(C_1 + C_2) + \sqrt{(C_1 + C_2)^2 - 4C_1C_2(1 - C_p^{-2})}}{2} \tag{9.12}$$

$$\text{where } C_p = \frac{1}{\sqrt{Q_1 Q_2}}$$

Here, Q_1 and Q_2 are the quality factors of the two resonant circuits. With the help of Eq. (9.10) S-parameter of antenna can be calculated. Hence, the radiation pattern of antenna can be calculated [14–20].

9.3   Specifications

Table 9.1 shows the design specifications of proposed antenna such as relative permittivity of substrate, feed point, dimension of patch and substrate.

Table 9.1 Specifications of antenna.

Substrate	Bakelite
Relative permittivity of substrate (ε_r)	4.78
Feed (x_0, y_0)	(5.2 mm, 9 mm)
Radius of patch (R) 1	5.0 mm
Thickness of substrate (h)	1.6 mm
Width (W_s) of the slot	1.0 mm
Length (L_n), Width (W_n) of notch	9.0 mm, 1.0 mm
Length of the slot (L_s)	7.0 mm

9.4   Result and Discussions

The measured data of the return loss along with the frequency of designed antenna is shown in Figure 9.7 using HFSS [21–23]. From the figure, it is seen that antenna operates at the frequency of 2.42, 3.26, 4.55 and 10.43

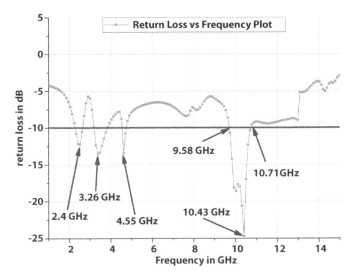

Figure 9.7 Plot of return loss vs frequency of designed antenna.

GHz with the percentage bandwidth of 21.59, 13.68, 4.42 and 11.14% respectively.

From Figure 9.8 it is obvious that both theoretical and simulated results for return loss vs frequency are in well agreement. The radiation pattern of

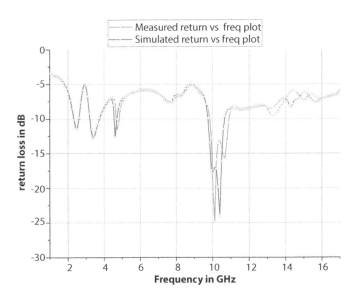

Figure 9.8 Measured, simulated plot of return loss versus frequency of antenna.

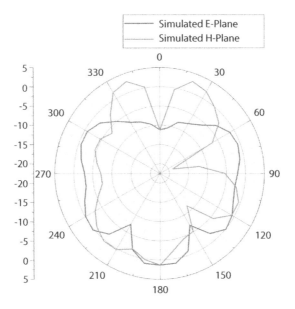

Figure 9.9 The plot of simulated E-plane, H-plane of radiation of antenna.

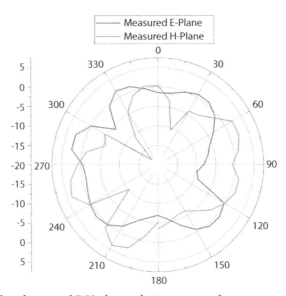

Figure 9.10 Plot of measured E/H-plane radiation pattern of antenna.

simulated as well as measured radiation pattern is shown in Figures 9.9 and 9.10 for E/H-Plane.

9.5   Conclusion

A parametrical study of multiband semicircular disk patch antenna has been presented in detail by measured and theoretical result which is compared with simulated result. The multiband operation for S/X-band applications is clear from the analysis and it may be concluded here that pair of slots loaded patch operate at multi resonance frequencies 2.42, 3.26, 4.55 and 10.43 GHz with the percentage bandwidth of 21.59, 13.68, 4.42 and 11.14% respectively.

References

1. Mishra, A., Singh, P., Ansari, J.A., Compact shorted microstrip patch antenna for dualband operation. *Prog. Electromagn. Res. C*, 9, 171–182, 2009, doi: 10.2528/PIERC09071007.
2. Krishna, D.D., Gopikrishna, M., Aanandan, C.K., Mohanan, P., Vasudevan, K., Compact dualband slot loaded circular disk Microstrip slot loaded antenna. *Prog. Electromagn. Res.*, 63, 245–255, 2008.
3. Ray, K.P. and Krishna, D.D., Compact dualband suspended semicircular microstrip patch antenna with half U-slot. *Microw. Opt. Technol. Lett.*, 48, 2021–2024, 2006.
4. Ansari, J.A., Mishraand, A., Vishvakarma, B.R., Half U-slot semicircular disk patch antenna for GSM mobile phones and optical communications. *Prog. Electromagn. Res. C*, 18, 31–45, 2011.
5. Ansari, J.A., Mishra, A., Vishvakarma, B.R., Analysis of L-shaped slot loaded circular disk patch antenna for satellite and radio telecommunication. *Int. J. Wirel. Pers. Commun.*, 70, 927–943, 2013.
6. Ansari, J.A., Mishra, A., Yadava, N.P., Singh, P., Vishvakarma, B.R., Analysis of W-slot loaded patch antenna for dualband operation. *Int. J. Electron. Commun.*, 66, 32–38, 2012.
7. Shivnarayan and B. R. Vishvakarma, Analysis of inclined slot loaded patch for dualband operation. *Microw. Opti. Techno Lett.*, 48, 2436–2441, 2006.
8. Tiang, J.J., Islam, M.T., Misran, N., Mandeep, J.S., Circular Microstrip antenna for dualband frequency RFID application. *Prog. Electromagn. Res.*, 120, 499–512, 2011.
9. Chen, L.C. *et al.*, Resonant frequency of circular disk printed circuit antenna. *IEEE Trans. Antennas Propag.*, 25, 595–596, 1997.

10. Garg, R., Bhartia, P., Bhal, I., Ittipiboon, *Microstrip Antenna Design Hand Book*, Artech House, Boston, 2003.
11. Kumer., G. and Ray, K.P., *Broadband Microstrip Antennas*, pp. 74–79, Artech House, Boston, 2003.
12. Zhang, X.X. and Yang, F.N., Study of slit cut on microstrip antenna and its application. *Microw. Opt. Technol. Lett.*, 18, 297–300, 1998.
13. Bahal, I.J., *Lumped elements for RF and microwave circuits*, p. 187, Artech House, Boston, 2003.
14. Balanis, C.A., *Antenna Theory analysis and design*, second edition, Wiley, New York, 1997.
15. Meshram, M.K. and Vishvakarma, B.R., Gap-coupled microstrip array antenna for wide band operation. *Int. J. Electron.*, 88, 1161–1175, 2001.
16. Kumar, A., Albreem, M.A., Gupta, M., Alsharif, M.H., Kim, S., Future 5G Network Based Smart Hospitals: Hybrid Detection Technique for Latency Improvement. *IEEE Access*, 8, 153240–153249, 2020.
17. Kumar, A., Gupta, M., Le, D.N., Aly, A.A., PTS-PAPR Reduction Technique for 5G Advanced Waveforms Using BFO Algorithm. *Intell. Autom. Soft Comput.*, 27, 3, 713–722, 2021.
18. Meena, K., Gupta, M., Kumar, A., Analysis of UWB Indoor and Outdoor Channel Propagation. *2020 IEEE International Women in Engineering (WIE) Conference on Electrical and Computer Engineering (WIECON-ECE)*, IEEE, pp. 352–355, 2020.
19. Gupta, M., Chand, L., Pareek, M., Power preservation in OFDM using selected mapping (SLM). *Int. J. Stat. Manage. Syst.*, 22, 4, 763–771, 2019.
20. Wolf, E.A., *Antenna Analysis*, Artech House, Narwood MA, USA, 1988.
21. Ansys software, Inc., HFSS simulation software, version 18, 2019.
22. Siddiqui, M.G., Saroj, A.K., Tiwari, D., Ansari, J.A., Multi-band fractaled triangular microstrip antenna for wireless applications. *Prog. Electromagn. Res. M*, 65, 51–60, 2018.
23. Saroj, A., Siddiqui, M.G., Kumar, M., Ansari, J.A., Design of multiband quad-rectangular shaped microstrip antenna for wireless applications. *Prog. Electromagn. Res. M.*, 59, 213–221, 2017.

10

A Rectangular Microstrip Patch Antenna with Defected Ground for UWB Application

Suraj Kumar[1], Arun Kumar[1]*, Manoj Gupta[1], Kanchan Sengar[1], Mohit Kumar Sharma[1] and Manisha Gupta[2]

[1]Department of Electronics and Communication, JECRC University, Jaipur, India
[2]Department of Physics, University of Rajasthan, Jaipur, India

Abstract

Microstrip receiving wires with unidirectional radiation examples and stable increases are the most valuable for this reason. To upgrade the data transfer capacity of a Microstrip receiving wire various advancements are utilized by energetic scientists that incorporate the utilization of Metamaterial. The basic structures with steady and moderate addition and huge data transfer capacity make a Microstrip reception apparatus appropriate for different applications in remote correspondence, for example, base station radio wires or wideband staged cluster receiving wires. Microstrip reception apparatuses display multiband attributes by utilizing metamaterials which are falsely designed materials. In this study, we introduce a rectangular microstrip slotted antenna with the absconded ground for UWB operations. The proposed rectangular microstrip patch antenna is comprised of FR-4 substrate with dielectric constant ($\varepsilon r = 4.4$) of measurement 20 mm × 30 mm. The reproduced outcomes uncover that the execution of the recommended antenna is phenomenal with return loss = −55.6 dB, gain = 2.311 dB & directivity = 3.698 dBi at 5.277 GHz & with return loss = −18.145, gain = 1.991 dB & directivity = 2.884 dBi at 3.646 GHz. We accomplished 2.7814 GHz bandwidth which lays in the middle of 3.223 to 6.0044 GHz band.

Keywords: UWB, microstrip antenna, return loss

*Corresponding author: arun.kumar1986@live.com

Arun Kumar, Manoj Gupta, Mahmoud A. Albreem, Dac-Binh Ha and Mohit Kumar Sharma (eds.) *Wearable and Neuronic Antennas for Medical and Wireless Applications*, (191–202) © 2022 Scrivener Publishing LLC

10.1   Introduction

In the ongoing years, remote interchanges frameworks have developed up quickly. These days, frameworks like portable correspondence, remote sensor systems are broadly utilized. Numerous strategies are utilized to incorporate a portion of these remote frameworks into one station. High selectivity channels are utilized to diminish the contortion worried to the obstruction starting with one help then onto the next assistance and accomplished disengagement between both the sources of info. By utilizing the multi-recurrence radio wire, an incredible improvement in front-end can be accomplished. In remote correspondence, the Micro-strip antenna is generally utilized. Multi-recurrence can be acquired by including a parasitic component of various thunderous recurrences. Minimal multi-recurrence reception apparatus with low recurrence can be planned by utilizing metamaterials. Without expanding the size of the single aerial double mode fix reception apparatus is gotten by utilizing radio wire and metamaterial collectively. Today, it is very well understood that antennas play an important function in a wireless communication system. With the increase of demands in broadband communication, we need compact size antennas with ease in fabrication and easy integration with microwave integrated circuits, which are fulfilled by microstrip patch antennas. But at the same time, it suffers some disadvantages such as low gain and narrow bandwidth [1, 2]. The researchers introduced planned and researched a few kinds of taking care of exhibitions of microstrip fix flying. The results of the test uncover that the recommended ethereal with the Inset Feed has advanced the boundaries at the recurrence of 1.6 GHz [3]. Taking care of procedure greatly affects the reception apparatus structure. It chooses the level of opportunity, returns misfortune, and subsequently the addition of the ideal radiation framework. A fixed radio wire can be planned to utilize different taking care of strategies. Rather than utilizing direct contact taking care of, if vicinity coupled taking care of is utilized, a bigger transfer speed is yielded with no corruption in the front-to-back proportion of the radio wire [4]. The creators projected a plan of microstrip reception apparatus devising a unique arrangement utilizing the fractional ground method. The anticipated aeronautical is apposite for several band operations. It includes fractional ground flat and overhauled circular smudge with glass epoxy substrate significant. The reproduced outcomes indicated that the proposed receiving wire has return loss (R ≤12 dB), high increase, and multi groups at the diverse resounding frequencies. The recommended airborne is of packed measurement which makes it reasonable for compact gadgets. The plan of the arranged ethereal can be altered by just expanding the region deprived of any adjustment in the proportions of the reception apparatus which is a significant concern in regards to configuration purposes [5]. The creators proposed a structure for a coplanar that took care of repeated flying taking CSRRs with proportionate structure and attached relatively over the maximum sheet of the reception apparatus for secluded communication structures. Recreation effects illustrate that the recurrence of the elevated framework can be portrayed by the structural boundaries of the CSRR. The projected receiving wire framework can be used for various remote frameworks by basically shifting the structure of CSRR [6]. The researchers introduced planned and researched a few kinds of taking care of exhibitions of microstrip fix flying. The results of the test uncover that the recommended ethereal with the Inset Feed has advanced the boundaries at the recurrence of 1.6 GHz. Taking care of procedure greatly affects the reception apparatus structure. It chooses the level of opportunity, returns misfortune, and subsequently the addition of the ideal radiation framework. A fixed radio wire can be planned to utilize different taking care of strategies. Rather than utilizing direct contact taking care of, if vicinity coupled taking care of is utilized, a bigger transfer speed is yielded with no corruption in the front-to-back proportion of the radio wire [7]. The examination work of plan, recreation, improvement, and estimation is done on the rectangular microstrip fix receiving wire for the ISM band of frequencies is introduced in this [8]. This rectangular fix radio wire is taken care of by a microstrip line feed. Its reverberation recurrence execution is concentrated between 2.4 and 2.5 GHz. In [8], an antenna is projected to expand the execution of the traditional arrangement took care of dipole pair radio wire by including a split-ring resonator. Right off the bat, an ordinary SDP reception apparatus that capacities in the recurrence scope of 1.7–2.3 GHz is planned by modifying the separations of the dipoles and the inter planetary concerning the two fold introductions. At that point, the impacts of SRR on the impedance, data transfer capacity, and addition are inspected. The trial effects indicate that the projected radio wire has an expanded increase. In [9], proposed a plan for a coordinating parting ring resonator utilized with microstrip ethereal to improve increase and data transfer capacity esteems. The experimental outcomes reveal that the resounding highlights of the microstrip ethereal are misrepresented by the span of the system, the width of the ring, and the quantity of spaces. The projected radio wire framework reverberates at subordinate reverberation frequencies. It is identified that the foreseen flying is humble to execute and shows a consistent radiation plan and great extension. The scratching of CSRR on rectangular fix gathering mechanical assembly is used for downsizing of a conventional fix accepting wire. The organized indistinguishable circuit is used to design multiband radio wires and have versatile assignments for remote correspondence structures. The essential resonation is spoken to by the semi-static resonation of the CSRR while the resulting resonation is started by the rectangular fix. Incredible comprehension among imitated and evaluated results is gained. The practically identical circuit model endorsement for the proposed arrangement gives a fundamental and away from for the structure of this sort of twofold band radio wires, allowing theory toward multiband getting wire arrangements of insignificant exertion and straightforward participating later on multisystem handsets [10]. As ordinary receiving wires are a cumbersome and exorbitant piece of an electronic framework, innovation of coordinated circuit advancement makes Microstrip reception apparatus extremely reduced and monetarily reasonable [11]. For additional upgrade in the exhibition of little measured radio wires in remote correspondence. A couple of Split Ring Resonator (SRR) executives are affixed over the primary dipole for exceptional yield misfortune activity. Reception apparatus execution is produced by the separation among the dual executives. The impact of SRR is broke down by estimating the information and acknowledged addition attributes. Through utilizing metamaterial over the structure of the SDP receiving wire, generally wide data transfer capacity and stable increase are accomplished when contrasted with that of QY radio wires. The recreation outcomes for the projected receiving wire spread 802.11 gauges for Wireless applications and can likewise be helpful for the 802.15 norms [12]. A microstrip shortened UWB radio wire is executed in the assortment 5.2–5.8 GHz. The recommended arrangement encompasses roundabout shortened and T-formed spaces for band functioning. The ideal outcomes can be acquired by choosing the receiving wire boundaries. Favorable circumstances of the proposed reception apparatus incorporate little dimension, and basic structure [13–15]. In this proposed antenna, we attempted to surmount the problem of bandwidth. The construction of a microstrip aerial comprises of a radiating module on the dielectric substrate. The survey and conception of the proposed antenna is shown in the research report. In our antenna design, we have inserted rectangular slots at the glowing element. In the present invention, the role of a limited ground level promoted the extension of the spectrum [16–18]. The first thing, we evaluated the results that we were getting in the initial ground plane. A limited ground smooth permitted us to increase the spectrum. Thereafter, we varied the measurements (span, thickness) of the spaces implanted in the fractional ground level, which likewise permitted us to spread out the spectrum. After sustained changes in the proportions of the slots we achieve an antenna that has wider bandwidth and a broader frequency range. Proposed antenna can be used in satellite communication, Wi-Fi devices, LTE, WI-Max, ISM, Indoor UWB system and wireless computer networks [19–23]. In this work, the rectangular UWB antenna is considered and projected. The rectangular UWB antenna is notched. The various parameters of antenna design are considered. The CST simulation results of s-parameters, gain, directivity and VSWR of the simulated aerial are analyzed.

10.2 Antenna Design

To defeat the contracted spectrum problem of the patch aerial, we have attempted to put together two pieces which have different resonance characteristics. Annular Ring antenna and defected ground was a good beginning point because its multiband characteristics. Likewise, implementation of miniaturized shapes is giving an adjusting ability on the transmitting aerial. In the taking care of system, two special substrates are utilized that is segregated by ground plane having a hole. These structures can give great outcomes. We used CST Studio Suit 2018 for simulations. After parametric simulation, we have gathered a band from 4.96 to 14.55 GHz in 30 × 30 mm² plate.

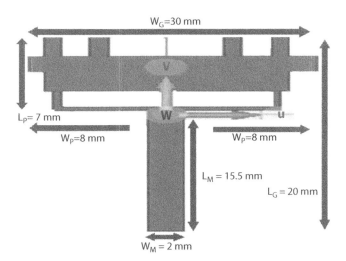

Figure 10.1 Main antenna geometry.

Figure 10.1 indicates the observation of the rectangular patch antenna for UWB operation. The projected aerial is analyzed with different parameters to astounded the disadvantage of constricted spectrum. The aerial is designed with slotted rectangular radiating patch.

Figure 10.2 displays the upper and lowest structure of the proposed UWB aerial. The aerial is intended on an FR4 of 20×30 mm^2 with a $\varepsilon_r = 4.4$. The depth of the substrate is H = 1.59 mm. The specification of proposed

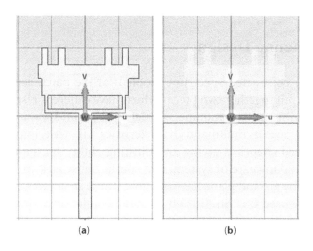

(a)   (b)

Figure 10.2 (a) Uppersight, and (b) lowestsight.

Table 10.1 Parameters of the proposed antenna.

Parameters	Dimensions
Ground plane	$L_G = 20$ mm; $W_G = 30$ mm
Substrate	$L_S = 20$ mm; $W_S = 30$ mm
Patch	$L_P = 7$ mm; $W_P = 16$ mm
Microstrip line	$L_M = 15.5$ mm; $W_M = 2$ mm
Slots	$S_{11} = 1.5$ mm; $S_{1w} = 2.5$ mm $S_{21} = 1.5$ mm; $S_{2w} = 2.5$ mm $S_{31} = 1.5$ mm; $S_{3w} = 2.5$ mm $S_{41} = 1.5$ mm; $S_{4w} = 2.5$ mm $S_{51} = 2.5$ mm; $S_{5w} = 6$ mm $S_{61} = 2.5$ mm; $S_{6w} = 1.5$ mm $S_{71} = 2.5$ mm; $S_{7w} = 1.5$ mm $S_{81} = 2$ mm; $S_{8w} = 1$ mm

aerial is made in Table 10.1. Aerial feeding is achieved by microstrip line in directive to acclimate to 50 Ω. We have done a parametric analysis to finalize the parameters of the projected aerial. The different limitations of the projected aerial are analyzed to get good results.

10.3   Simulation Results

S-parameter spectacles the forfeiture of power in the signal imitated owing to the cut-off in the communication route. This is the supreme serious consequence of reproduction as it displays the recurrence on which radio wire is resounding. The estimation of return loss requisite not be certain. It very well may be acceptable as the progressively negative worth is accomplished. Figures 10.3 and 10.4, show that the return loss of the projected UWB aerial through and deprived of notching. The return loss without notching and with notching is −22 dB at 3.6 GHz, −36 dB at 5.263 GHz and −18 dB at 3.6 GHz, −55 dB at 5.263 GHz. Figure 10.5 demonstrates the simulated result of VSWR of the proposed antenna. The ranges of VSWR vary between 1 and 2 dB in the spectrum of 2.4451 to 7.959 GHz. The proposed UWB antenna shows excellent VSWR characteristics in the spectrum scope. The VSWR of the suggested antenna is 1.6153. The majority of the aerials are worked in the far-field area. This is where the radiation design doesn't alter the character with separation. The far-field design contributes data about the directivity and addition of the reception apparatus. Their qualities must be sure of better outcomes. Figure 10.6 presents the 3-dimensional radiation patter of the proposed antenna. The far field design gives data about the directivity and addition of the receiving wire. The gain of the antenna is 1.99 dB at 3.6 GHz and 2.31 dB at 5.2 GHz.

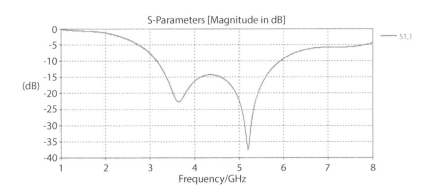

Figure 10.3 S-parameter (No notching).

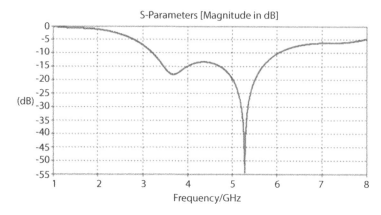

Figure 10.4 S-parameter for different slots dimension.

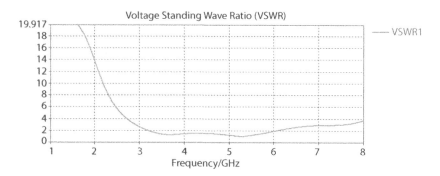

Figure 10.5 VSWR.

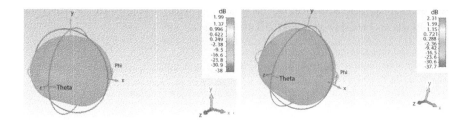

Figure 10.6 3-D radiation patterns.

10.4 Conclusion

The projected antenna is designed by CST Microwave Studio Simulator. Several parameters are changed to obtain the wanted effects. In the proposed plan, the role of a partial ground plane promoted the expansion of

Table 10.2 Comparison of different antennas.

Ref.	Design	Return loss	Size	Gain
[15]	Rectangular Patch Antenna for Dual Band Application	−30.42 dB	60 × 70 mm²	3.4 dB
[16]	Band-notch UWB Antenna	−33 dB	16 × 30 × 1.6 mm²	3.1 dB
[17]	UWB Antenna with Molded Interconnect device Technology	−19.5 dB	10 × 10 × 0.2 mm	4.2 dB
[18]	Printed UWB Monopole Antenna	–	20 × 18 mm²	3.8 dB
[19]	Notched UWB MIMO-Antenna	−22 dB	18 × 34 mm²	3.1 dB
	Proposed Antenna	−55.6 dB	20 × 30 mm²	2.311 dB

the spectrum. The first thing, we evaluated were the results that we were getting from the initial ground plane so as to enlarge the bandwidth. Thenceforth, we varied the different parameters of the antenna and slots inserted in the partial ground plane. After sustained changes in the proportions of the slots, we reach an antenna that has wider bandwidth and a broader frequency range. The simulated results reveal that the execution of the proposed antenna is excellent with Return loss = −55.6 dB, Gain = 2.311 dB & directivity = 3.698 dB at 5.277 GHz & with Return loss = −18.145, Gain = 1.991 dB & directivity = 2.884 dBi at 3.646 GHz. We are getting return loss at 5.227 & 3.646 GHz respectively, which covers some part of C band. We achieved 2.7814 GHz bandwidth which lies in between 3.223 and 6.0044 GHz band. Proposed antenna can be used in satellite communication, Wi-Fi devices, LTE, WI-Max, ISM, Indoor UWB system and wireless computer networks. The past and previous studied are given in Table 10.2.

References

1. Kumar, A. and Singh, M.K., Band-Notched Planar UWB Microstrip Antenna with T-Shaped Slot. *Radioelectron. Commun. Syst.*, 61, 371, 2018. https://doi.org/10.3103/S0735272718080058.

2. Ojaroudi, M. and Ojaroudi, N., Ultra-Wideband Small Rectangular Slot Antenna With Variable Band-Stop Function. *IEEE Trans. Antennas Propag.*, 62, 1, 490–494, Jan. 2014.

3. Arora, A., Khemchandani, A., Rawat, Y., Singhai, S., Chaitanya, G., Comparative study of different feeding techniques for rectangular microstrip patch antenna. *IJIREEICE*, 3, 5, 32–5, 2015.

4. McGregor, E.A.S. and Ian, ElgaidKhaled, 200 GHz Broadband Proximity Coupled Patch Antenna. *IEEE International Conference on Ultra-Wideband (ICUWB-2009)*, pp. 404–407, 2009.

5. Singh, V.K. and Naresh, B., Multi resonant microstrip antenna with partial ground for radar application. *J. Telecommun. Switching Syst. Netw.*, 2, 1, 01–05, 2015.

6. Aznabet, M., El Mrabet, O., Floc'h, J.M., Falcone, F., Drissi, M.H., A coplanar waveguide-fed printed antenna with complementary split ring resonator for wireless communication systems. *Waves Random Complex Media*, 25, 1, 43–51, 2015.

7. Pandeeswari, R. and Raghavan, S., Microstrip antenna with complementary split ring resonator loaded ground plane for gain enhancement. *Microw. Opt. Technol. Lett.*, 57, 2, 292–296, 2015.

8. Ortiz, N., Iriarte, J.C., Crespo, G., Falcone, F., Design and implementation of dual-band antennas based on a complementary split ring resonators. *Waves Random Complex Media*, 25, 3, 309–322, 2015.

9. Yeo, J., Lee, J.I., Park, J.T., Broadband series-fed dipole pair antenna with parasitic strip pair director. *Prog. Electromagn. Res. C*, 45, 1–13, 2013.

10. Kumar, A. and Choudhary, M., Dual Band Modified Split-Ring Resonator Microstrip Antenna for Wireless Applications. *Natl. Acad. Sci. Lett.*, 43, 237–240, 2020. https://doi.org/10.1007/s40009-019-00845-7.

11. Lin, Y.-C. and Hung, K.-J., Compact Ultrawideband rectangular aperture antenna and band-notched designs. *IEEE Trans. Antennas Propag.*, 54, 11, 3075–3081, Nov. 2006.

12. Emadian, S.R., Ghobadi, C., Nourinia, J., A novel compact dual band-notched slot antenna for ultrawideband applications. *Microw. Opt. Technol. Lett.*, 52, 1365–1368, 2012.

13. Peixeiro, C., Microstrip Antenna Papers in the IEEE Transactions on Antennas and Propagation [EurAAP Corner]. *IEEE Antennas Propag. Mag.*, 54, 1, 264–268, Feb. 2012. doi: 10.1109/MAP.2012.6202561.

14. Munson, R.E., Conformal Microstrip Antennas and Microstrip Phased Arrays. *IEEE Trans. Antennas Propag.*, AP-22, 74–78, January 1974.a.

15. Karli, R. and Ammor, H., Rectangular Patch Antenna for Dual-Band RFID and WLAN Applications. *Wirel. Pers. Commun.*, 83, 2, 995–1007, 2015.

16. Duy, T.N. and Van, Y.V., Combining two methods to enhance band-notch characteristic of ultra wide band antenna. *2009 International Conference on Advanced Technologies for Communications*, HaiPhong, pp. 206–210, 2009.

17. Unnikrishnan, D., Kaddour, D., Tedjini, S., Bihar, E., Saadaoui, M., CPW-Fed Inkjet Printed UWB Antenna on ABS-PC for Integration in Molded Interconnect Devices Technology. *IEEE Antennas Wirel. Propag. Lett.*, 14, 1125–1128, 2015.

18. Abdollahvand, M., Dadashzadeh, G., Mostafa, D., Compact Dual Band-Notched Printed Monopole Antenna for UWB Application. *IEEE Antennas Wirel. Propag. Lett.*, 9, 1148–1151, 2010.

19. Chandel, R., Gautam, A.K., Rambabu, K., Tapered Fed Compact UWB MIMO-Diversity Antenna With Dual Band-Notched Characteristics. *IEEE Trans. Antennas Propag.*, 66, 4, 1677–1684, April 2018.

20. Kumar, A., Albreem, M.A., Gupta, M., Alsharif, M.H., Kim, S., Future 5G Network Based Smart Hospitals: Hybrid Detection Technique for Latency Improvement. *IEEE Access*, 8, 153240–153249, 2020.

21. Kumar, A., Gupta, M., Le, D.N., Aly, A.A., PTS-PAPR Reduction Technique for 5G Advanced Waveforms Using BFO Algorithm. *Intell. Autom. Soft Comput.*, 27, 3, 713–722, 2021.

22. Meena, K., Gupta, M., Kumar, A., Analysis of UWB Indoor and Outdoor Channel Propagation. *2020 IEEE International Women in Engineering (WIE) Conference on Electrical and Computer Engineering (WIECON-ECE)*, IEEE, pp. 352–355, 2020.

23. Gupta, M., Chand, L., Pareek, M., Power preservation in OFDM using selected mapping (SLM). *J. Stat. Manage. Syst.*, 22, 4, 763–771, 2019.

11

Waveform Optimization in Multi-Carrier Communications for 5G Technology

**Muhammad Moinuddin[1,2]*, Ubaid M. Al-Saggaf[1,2]
and Jawwad Ahmed[3]**

*[1]Center of Excellence in Intelligent Engineering Systems (CEIES), King Abdulaziz
University, Jeddah, Saudi Arabia
[2]Electrical and Computer Engineering Department, King Abdulaziz University,
Jeddah, Saudi Arabia
[3]Electrical Engineering Department, Usman Institute of Technology,
Karachi, Pakistan*

Abstract

With respect to 4G, future 5G radio communication systems and beyond are expected to bring new applications and services, with challenging requirements in terms of quality of service, spectrum efficiency, latency, and resilience to extreme, yet artificial, time and frequency dispersion impairments [1–7]. Tactile Internet and next-generation connected autonomous or driver-assist cars are among the plethora of expected 5G applications that require very stringent latencies. Multimedia Broadcast Multicast Services (MBMS) are evolved forms of Single-Frequency Networks (SFN), first used in Digital Video Broadcasting-Terrestrial (DVB-T) systems, will also be among the applications offered by 5G. Moreover, Cloud-Radio-Access-Network (C-RAN) based networks are expected to be among the enhancements that will profitably be adopted in 5G. All these will bring a tremendous artificial increase in experienced channel dispersion in time and frequency. Due to these unprecedented artificial propagation impairments, waveform optimization is not anymore an option but a necessity for the guarantee of the best achievable 5G radio interface efficiency and quality of service.

Keywords: Multiuser communications, multicarrier communications, OFDM, FBMC, 5G, waveform optimization, outage probability, POPS

**Corresponding author*: mmsansari@kau.edu.sa

Arun Kumar, Manoj Gupta, Mahmoud A. Albreem, Dac-Binh Ha and Mohit Kumar Sharma (eds.)
Wearable and Neuronic Antennas for Medical and Wireless Applications, (203–216) © 2022
Scrivener Publishing LLC

11.1   Introduction

Multicarrier modulations (MC) are strong candidates for the 5G as they have better spectral efficiency and operational flexibility. The Cyclic-prefix orthogonal frequency division multiplexing (CP-OFDM) is the most popular technique in MC as it is easy to implement. However, the OFDM has poor spectral efficiency due to limitedfiltering options available.

Another candidate considered for multiple radio access in 5G is the Filter Bank Multi-Carrier (FBMC) [8–10]. The FBMC is a modified version of OFDM, where the rectangular transmit and receive waveforms are replaced by any other more frequency localized pulse shapes, which offer a more confined spectrum, relevant for spectrum-sharing scenarios. Frequency localization, however, comes at the cost of longer filter lengths, inflicting a hard blow on low latency requirements. The main drawback of FBMC is the use of empirical classes of parametric waveforms, sought for their good OOB emission reduction capabilities, with no guaranteed resilience to extreme channel dispersion impairments and receiver synchronization imperfections.

As used in 4G LTE, conventional OFDM will not alone be able to handle the various and sometimes contradictory requirements of 5G applications, even if several CP durations are considered. Waveform design will enable the optimization of the waveform codebook of each 5G application, depending on the corresponding propagation impairments' statistics and QoS requirements, latency, and coexistence with other applications. Moreover, waveform design will also enable the optimization of the transmission and reception waveforms as well as the underlying time-frequency lattice layout for prescribed quality of service requirements.

11.2   Related Literature Review

The FBMC modulation is a recent type of MC technique that is developed to improve the performance of the conventional OFDM by minimizing Intersymbol interference (ISI) and inter-carrier interference (ICI) [8, 9].

The task of waveform optimization is not recent. It has started around in the mid of the nineties (1995), with the introduction of the Isotropic Orthogonal Transform Algorithm (IOTA) waveform used in OFDM-OQAM systems [11]. Recently, a new technique for the fast optimization of multicarrier waveforms is developed [12]. Due to the ping-pong nature of the underlying offline optimization process, which switches between the optimization tasks of transmit and receive pulses, this technique was called Ping-pong Optimized Pulse Shaping (POPS) [12–15].

So far, work on POPS enabled us to optimize both transmit and receive waveforms for isolated communications. Whereby all impairments related to radio channel propagation dispersions, synchronization mismatch, and misalignment in time and frequency, are taken into account in the maximization of the achievable Signal-to-Interference and Noise Ratio (SINR) [15–19]. This optimization has mainly been accomplished for the OFDM and the FBMC systems, with underlying rectangular [20] and hexagonal time-frequency lattices [21–23].

11.3   System Model: OFDM System

The transmission system of the OFDM system is illustrated in Figure 11.1. It can be observed from Figure 11.1 that the OFDM transmitter consists first of a modulator block, which maps the information bits to modulated symbols. This is followed by a serial to parallel converter whose output is fed to the IFFT block, converting data to the time domain. This data is again converted to serial data before transmission. In the case of CP-OFDM, the cyclic prefix is added to the serial data, which is then transformed into an analog format before transmission through the communication channel using Digital-to-Analog (DAC) block. The receiver of the OFDM uses the mirror image of the transmission blocks.

The transmitted basis pulse for the lth sub-carrier and kth symbol is defined as:

$$g_{l,k}(t) = p(t - kT)e^{\frac{j2\pi}{F(t-kT)}}e^{j\theta_{l,k}} \qquad (11.1)$$

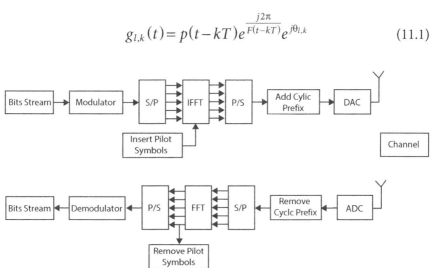

Figure 11.1 OFDM transmitter and receiver systems.

where is the prototype filter, F is the spacing between subcarrier, and T is spacing in time. The sampling rate is evaluated by $f_s = 1/\Delta t = FN_{FFT}$, where $N_{FFT} \geq L$. Here, N_{FFT} is used to show the FFT size. The sampled basis pulse $g_{l,k}(t)$, can be expressed as vector $g_{l,k} \in C^{N \times 1}$, given by

$$\left[g_{l,k} \right]_i = \sqrt{\,t\,}\, g_{l,k}(t) \Big|_{t=(i-1)\ t-\frac{OT_o}{2}+T} \quad for\ i = 1,\ \ldots,\ N \qquad (11.2)$$

with

$$N = OT_0 + T\,(k-1)\,f_s \qquad (11.3)$$

Next, by combining all the vectors in a matrix $G \in C^{N \times LK}$, given by

$$G = \left[g_{1,1} \cdots\ g_{L,1} g_{1,2} \cdots\ g_{L,k} \right] \qquad (11.4)$$

and by representing the combined data vector for all the transmitted symbols as vector $x \in C^{LK \times 1}$,

$$x = vec \left\{ \begin{bmatrix} x_{1,1} & \cdots & x_{1,k} \\ \vdots & \ddots & \vdots \\ x_{L,1} & \cdots & x_{L,k} \end{bmatrix} \right\} \qquad (11.5)$$

$$x = \left[x_{1,1} \cdots\ x_{1,k} x_{L,1} \cdots\cdots\ x_{L,k} \right]^T \qquad (11.6)$$

We can write the combined sampled transmit signal $S \in C^{N \times 1}$ as

$$S = Gx \qquad (11.7)$$

The impulse response under multipath fading in a time-variant channel can be expressed as matrix $H \in C^{N \times N}$, whose entries are

$$[H]_{i,j} = h[i-j,i] \qquad (11.8)$$

In the OFDM, the receiver matrix is different from the transmitter. Thus, if the receiver matrix is represented by Q, the received symbols can be expressed as

$$y = Q^H r = Q^H H G x + n \tag{11.9}$$

where n represents the additive noise term which is usually modeled as a Gaussian random variable.

11.4 POPS: A Popular Existing Method for OFDM Waveform Optimization

As POPS is one of the popular existing methods for waveform optimization in OFDM systems, we provide its fundamental concept. According to the POPS method, waveform optimization is done by maximizing the SINR expression. To do so, the received SINR of the OFDM system is expressed as a generalized Rayleigh quotient on the transmission waveform vector φ, given the receiving waveform vector Ψ, and vice-versa. Thus, the baseband transmitted signal vector of the OFDM can be written as:

$$\mathbf{e} = \sum_{mn} a_{mn}\, \varphi_{mn}, \tag{11.10}$$

where ∂_{mn} is the symbol transmitted at "nT" time and "mF" frequency, φ_{mn} is the transmission waveform vector φ after having a time shift of nT and a frequency shift of mF. This signal vector is digital-to-analog converted, analog-filtered, and then transposed to the carrier band for effective transmission. The resulting base-band received vector is given by:

$$\mathbf{r} = \sum_{mn} a_{mn}\, \tilde{\varphi}_{mn} + \mathbf{n}, \tag{11.11}$$

where $\tilde{\varphi}_{mn}$ is the vector φ_{mn} after including channel distortion effects and $\mathbf{n}$ is the AWGN. In order to make a decision,the received signal is projected on the corresponding waveform vector as

$$\Lambda_{kl} = \langle \Psi_{kl}, \mathbf{r} \rangle = \Psi_{kl}^H \mathbf{r}, \tag{11.12}$$

where Ψ_{kl} is the received vector Ψ with time and frequency shifted symbols, $(\cdot)^H$ is the Hermitian transpose operator, and $\langle \cdot, \cdot \rangle$ is the Hermitian scalar product. The decision variable on a_{kl} can be expanded as

$$\Lambda_{kl} = \underbrace{a_{kl}\left\langle \Psi_{kl}, \tilde{\phi}_{kl} \right\rangle}_{U_{kl}} + \underbrace{\sum_{(m,n)\neq(k,l)} a_{mn}\left\langle \Psi_{kl}, \tilde{\phi}_{mn} \right\rangle}_{I_{kl}} + \underbrace{\left\langle \Psi_{kl}, \mathbf{n} \right\rangle}_{N_{kl}}, \qquad (11.13)$$

with U_{kl}, I_{kl} and N_{kl} is the useful, the interference and the noise terms. The SINR, which is common to all transmitted symbols, is defined as the ratio of the average power of the useful term U_{kl} to the sum of the average powers of the interference term I_{kl} and the noise term N_{kl}. After some mathematical derivations, omitted here for lack of space, we show that this SINR can be set in one of two generalized Rayleigh quotient (GRQ) forms. The first form,

$$SINR = \frac{\Psi^{H}\mathbf{KS}^{\phi}_{S(p;v)}\Psi}{\Psi^{H}\mathbf{KIN}^{\phi}_{S(p;v)}\Psi}, \qquad (11.14)$$

is the ratio of receiving waveform Ψ dependent quadratic forms. Here, $\mathbf{KS}^{\phi}_{S(p;v)}$ and $\mathbf{KIN}^{\phi}_{S(p;v)}$ are the useful and interference plus noise Kernels, which are quadratic positive definite matrices, depending on the transmission waveform ϕ, the scattering function $S(p;v)$, and the signal-to-noise ratio SNR, and $\|\cdot\|$ is the Euclidian norm. This form is suitable for the optimization of the receiving waveform Ψ for a given transmission waveform ϕ. The optimum receiving waveform and corresponding SINR, obtained by the eig$(\cdot)$ function of MATLAB, are respectively the eigenvector and associated maximum eigenvalue of the GRQ. The second form,

$$SINR = \frac{\phi^{H}\mathbf{KS}^{\Psi}_{S(-p;-v)}\phi}{\phi^{H}\mathbf{KIN}^{\Psi}_{S(-p;-v)}\phi}, \qquad (11.15)$$

is the ratio of transmitting waveform ϕ dependent quadratic forms. Here $\mathbf{KS}^{\Psi}_{S(-p;-v)}$ and $\mathbf{KIN}^{\Psi}_{S(-p;-v)}$ being the useful and interference plus noise Kernels, which are quadratic positive definite matrices, depending on the receiving waveform Ψ and the scattering function $S(p;v)$. This second form is suitable for the optimization of the transmission waveform ϕ for a given receiving waveform Ψ.

The POPS algorithm [12], which is iterative, relies on alternate use of the two previous forms of SINR. As illustrated in Figure 11.2, each iteration is composed of two steps, namely a forward step involving the optimization of the receiving waveform, given the transmission waveform obtained in the previous iteration, and a backward step involving the optimization of the transmission waveform, given the optimized receiving waveform obtained

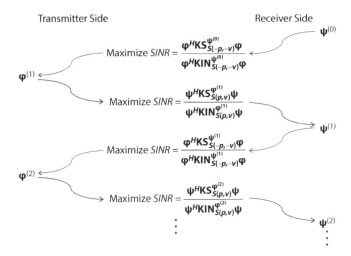

Figure 11.2 POPS philosophy of the waveforms. Optimization by the maximization of SINR.

in the forward step. The POPS algorithm inherits its name from this to-ing and fro-ing nature of the optimization process. The first iteration of the POPS algorithm starts with an arbitrary initial guess $\Psi^{(0)}$ of the reception waveform. In the first step of the i-th iteration, the POPS algorithm plugs the optimized receiving waveform $\Psi^{(i-1)}$, obtained at the end of the previous $(i-1)$-th iteration, in the Kernels of Eq. (11.14), to determine the best corresponding transmission waveform $\varphi^{(i)}$ at the i-th iteration. Then, in the second step, the POPS algorithm plugs the just obtained optimized transmission waveform $\varphi^{(i)}$ in the Kernels of Eq. (11.13) to determine the best corresponding receiving waveform $\Psi^{(i)}$ at the i-th iteration.

11.5 Proposed Method for the Waveform Optimization in OFDM Systems

In the proposed method, we propose to optimize the waveform by minimizing the outage probability. To do so, we first derive the outage probability of the OFDM symbol in closed form. The pqth symbol of the received signal given in Eq. (11.9) is

$$y_{pq} = q_{pq}^{H} H \hat{g}_{pq} + q_{pq}^{H} H \sum_{\substack{l'=0 \\ l' \neq p}}^{L-1} \sum_{K'=0}^{K-1} g_{l'K'} x_{l'K'} + q_{lK}^{H} H \sum_{\substack{K'=0 \\ K' \neq q}}^{K-1} g_{pq} x_{lK'} + q_{pq}^{H} n,$$

$$(11.16)$$

where q_{pq} and g_{pq} are the pqth column of the matrices Q and G, respectively. Next, in order to find expression for SINR, we calculated the energies of the terms in Equation (11.16) and using the assumption that $E|x_{pq}|^2 = 1$, for all p and q. Thus, the SINR for the pqth symbol can be expressed as

$$SINR_{pq} = \frac{\left|q_{pq}^H Hg_{pq}\right|^2}{\left|q_{pq}^H H\hat{g}_\alpha\right|^2 + \left|q_{pq}^H Hg_\beta\right|^2 + \sigma_n^2 Tr\left(q_{pq}q_{pq}^H\right)} \qquad (11.17)$$

where the notation $Tr( )$ is used to represent the trace operator. Here, the terms g_α and g_β represent, respectively,

$$g_\alpha = \sum_{\substack{l'=0 \\ l'\neq p}}^{L-1} \sum_{K'=0}^{K-1} g_{l'K'} \qquad (11.18)$$

and

$$g_\beta = \sum_{\substack{K'=0 \\ K'\neq q}}^{K-1} g_{pK'} \qquad (11.19)$$

Next, we employ the standard procedure for whitening transformation of the channel matrix, that is, $H = R_{R_X}^{1/2} \bar{H} R_{T_X}^{1/2}$. Now, by using the following definitions

$$q_{pq_R} = R_{R_X}^{1/2} q_{pq}^H \qquad \text{and} \qquad g_{pq_T} = R_{T_X}^{1/2} g_{pq}$$

we can rewrite the SINR in Eq. (11.17) as follows

$$SINR_{pq} = \frac{\left\|\bar{h}\right\|_{A_1}^2}{\left\|\bar{h}\right\|_{A_2}^2 + \sigma_n^2 Tr\left(q_{pq}q_{pq}^H\right)} \qquad (11.20)$$

where $\bar{h} = vec(\bar{H})$ is the vector obtained from the channel matrix $\bar{H}$. Here, the matrices A_1 and A_2 are defined as

$$A_1 = \left(q_{pqR} \otimes g_{pqT}\right)^H \left(q_{pqR} \otimes g_{pqT}\right) \tag{11.21}$$

$$A_2 = \left(q_{pqR} \otimes g_{\alpha T}\right)^H \left(q_{pqR} \otimes g_{\alpha T}\right) + \left(q_{pqR} \otimes g_{\beta T}\right)^H \left(q_{pqR} \otimes g_{\beta T}\right) \tag{11.22}$$

Thus, the outage probability of the pqth symbol can be evaluated as

$$P_{out\,p,q}\left(\gamma_{th}\right) = P\left(SINR_{pq} \leq \gamma_{th}\right) \tag{11.23}$$

which is equivalent to write as:

$$P_{out\,p,q}\left(\gamma_{th}\right) = P\left(\frac{\|\bar{h}\|_{A_1}^2}{\|\bar{h}\|_{A_2}^2 + \sigma_n^2\, Tr\left(q_{pq}q_{pq}^H\right)} \leq \gamma_{th}\right) \tag{11.24}$$

The probability appearing in the RHS of Eq. (11.24) can be evaluated using a recently developed approach of indefinite quadratic forms given in [19]. Thus, it can be shown that the evaluation of the above probability will give the following result:

$$P_{out\,p,q}\left(\gamma\right) = 1 - \sum_{n=1}^{Np} \left(\frac{\lambda_n^{Np}\, e^{-\frac{\gamma_{th}\sigma_n^2 Tr\left(q_{pq}q_{pq}^H\right)}{\lambda_n}}}{\prod_{\substack{i=1 \\ i \neq n}}^{Np} \left(\lambda_n - \lambda_i\right)|\lambda_n|}\, u\left(\frac{\gamma_{th}\sigma_n^2 Tr\left(q_{pq}q_{pq}^H\right)}{\lambda_n}\right)\right) \tag{11.25}$$

where $N = LK$, λ_n is the nth eigenvalue of the matrix $A1 - \lambda_{th}A2$, and $u()$ is the unit step function. Thus, the total outage probability of the time-frequency symbols will be:

$$P_{out}\left(\gamma\right) = \frac{1}{LK} \sum_{p=1}^{L} \sum_{q=1}^{K} P_{out\,p,q}\left(\gamma\right) \tag{11.26}$$

To obtain an optimized solution, we propose to minimize the following objective function

$$\min_{\{q_{00},q_{10},\ldots,q_{LK}\}} J\left(q_{00},q_{10},\ldots,q_{LK}\right) = P_{out}\left(\gamma\right) \tag{11.27}$$

11.6 Results and Discussion

In this section, we provide the results of the proposed waveform optimization developed in Section 11.5. In this context, the OFDM system is simulated for 24 subcarriers and 30 time symbols. In Figure 11.3, the expression of the total outage probability derived in Eq. (11.25) is validated via simulation results. It can be seen that the theoretical results are well-matched with the simulations one. Next, in Figure 11.4, the proposed waveform optimization method is implemented using Matlab built-in optimization function "*fmincon*" via the "Sequential Quadratic Programming" method. The results show that the proposed method minimizes the outage probability for all the threshold values showing improvement in overall outage probability.

11.7 Summary

In this chapter, we discussed the problem of waveform optimization in OFDM system. We propose to optimize the waveforms of the OFDM symbols by minimizing the total outage probability of all time-frequency symbols. For this task, we first derive a closed form expression for the outage probability of all the symbols. This is done by expressing the SINR in terms of the ratio of Indefinite Quadratic Forms. Thus, the outage probability is

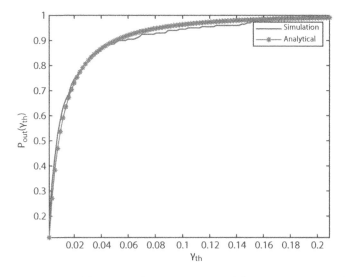

Figure 11.3 Validation of derived analytical outage probabilities via simulations theory.

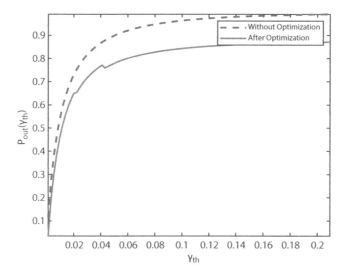

Figure 11.4 Outage probability performance of the proposed method.

evaluated by calculating the probability that the SINR is less than some given threshold value via employing the Fourier transform of a unit step function and multi-dimensional Gaussian integral solution. The derived analytical results are validated via Monte-Carlo simulations. The derived expression for the outage probability is then utilized to obtain the optimum solution for OFDM waveforms by minimizing it using Sequential Quadratic Programming, whose performance shows that the proposed method minimizes the total outage probability in obtaining the optimized waveforms.

References

1. Dahlman, E., Mildh, G., Parkvall, St., Peisa, J., Sachs, J., Selén, Y., Sköld, J., 5G wireless access: requirements and realization. *IEEE Commun. Mag.*, 52, 12, 2014.
2. Wunder, G., Kasparick, M., Wild, T., Schaich, F., Chen, Y., ten Brink, S. *et al.*, 5GNOW: Application Challenges and Initial Waveform Results. *Proceedings of Future Network & Mobile Summit*, July 2013.
3. Wunder, G., Kasparick, M., ten Brink, S., Schaich, F., Wild, T., Gaspar, I., Ohlmer, E., Krone, S., Michailow, N., Navarro, A., Fettweis, G., Ktenas, D., Berg, V., Dryjanski, M., Pietrzyk, S., Eged, B., 5GNOW: Challenging the LTE Design Paradigms of Orthogonality and Synchronicity. *Mobile and Wireless*

Communication Systems for 2020 and beyond (Workshop @ 77th Veh. Technol. Conf.: VTC2013-Spring), June 2013.

4. Kasparick, M., Wunder, G., Schaich, C.F., Wild, T., Berg, V., Cassiau, N., Dor, J., Ktnas, D., Dryjaski, M., Pietrzyk, S., Gaspar, I.S., Michailow, N., 5G waveform candidate selection,"Tech. Rep., D3.1 of 5G-Now, FP7 European Research Project, Nov. 2013.

5. Demestichas, P., Georgakopoulos, A., Karvounas, D., Tsagkaris, K., Stavroulaki, V., Lu, J., Xiong, C., Yao, J., 5G on the Horizon: Key Challenges for the Radio-Access Network. IEEE Veh. Technol. Mag., 8, 3, 2013.

6. Li, Q.C., Niu, H., Papathanassiou, A.T., Wu, G., 5G Network Capacity: Key Elements and Technologies. IEEE Veh. Technol. Mag., 9, 1, 2014.

7. Hossain, E. and Hasan, M., 5G cellular: key enabling technologies and research challenges. IEEE Instrumen.Meas. Mag., 183, 2015.

8. Farhang-Boroujeny, B., OFDM versus filter Bank multicarrier. IEEE Signal Process. Mag., 28, 3, 92–112, May 2011.

9. Nissel, R., , Rupp, M., Filter bank multicarrier modulation schemes for future mobile communications. IEEE J. Sel. Areas Commun., 35, 8, Schwarz August, 2017.

10. Zhao, Z., Lin, H., Siohan, P., Rajatheva, N., Luecken, V., Ishaque, A., FBMC-based air interface for 5G mobile: Challenges and proposed solutions. 9th International Conference on Cognitive Radio Oriented Wireless Networks and Communications (CROWNCOM), 2014.

11. Le Floch, B., Alard, M., Berrou, C., Coded orthogonal frequency division multiplex. Proceedings of the IEEE, vol. 83, pp. 982–996, 1995.

12. Siala, M., Abdelkefi, F., Hraiech, Z., Novel Algorithms for Optimal Waveforms Design in Multicarrier Systems. IEEE Wireless Commun.Networking Conf. (WCNC'2014), April 2014.

13. Hraiech, Z., Siala, M., Abdelkefi, F., Numerical Characterization for Optimal Designed Waveform to Multicarrier Systems in 5G. 22nd European Signal Processing Conference (EUSIPCO 2014, Lisbon, Portugal, pp. 1–5, September 2014.

14. Hraiech, Z., Abdelkefi, F., Siala, M., Characterization of ping-pong optimized pulse shaping-OFDM (POPS-OFDM) for 5G systems. IEEE Vehicular Technology Conference – Spring 2015 (VTC'S15), Glasgow, Scotland, May 2015.

15. Kumar, A., Albreem, M.A., Gupta, M., Alsharif, M.H., Kim, S., Future 5G Network Based Smart Hospitals: Hybrid Detection Technique for Latency Improvement. IEEE Access, 8, 153240–153249.

16. Kumar, A., Gupta, M., Le, D.N., Aly, A.A., PTS-PAPR Reduction Technique for 5G Advanced Waveforms Using BFO Algorithm. Intell. Autom. Soft Comput., 27, 3, 713–722, 2021.

17. Meena, K., Gupta, M., Kumar, A., Analysis of UWB Indoor and Outdoor Channel Propagation. 2020 IEEE International Women in Engineering (WIE) Conference on Electrical and Computer Engineering (WIECON-ECE), IEEE, pp. 352–355, 2020.

18. Gupta, M., Chand, L., Pareek, M., Power preservation in OFDM using selected mapping (SLM). *J. Stat. Manage. Syst.*, 22, 4.
19. Hraiech, Z., Abdelkefi, F., Siala, M., POPS-OFDM: Ping-Pong Optimized Pulse Shaping-OFDM for 5G Systems. *IEEE International Conference on Communications (ICC'15)*, London, UK, June 2015.
20. Hraiech, Z., Abdelkefi, F., Siala, M., POPS-OFDM with different Tx/Rx pulse shape durations for 5G systems. *5th International Conference on Communications and Networking (COMNET'2015)*, Hammamet, Tunisia, November 2015.
21. Ayadi, R., Kammoun, I., Siala, M., Efficient Offline Waveform Design Using Quincunx/Hexagonal Time-Frequency Lattices for 5G Systems, in: *Wireless Communications and Mobile Computing*, Hindawi, 2017.
22. Hraiech, Z., Abdelkefi, F., Siala, M., Zayani, R., Optimization of the PHYDYAS Waveforms Using the POPS Algorithm: POPS-PHYDYAS. *15th International Multi-Conference on Systems, Signals & Devices (SSD), Hammamet, 2018*, pp. 1067–1071, 2018.
23. Al-Naffouri, T.Y., Moinuddin, M., Ajeeb, N., Hassibi, B., Moustakas, A.L., On the Distribution of Indefinite Quadratic Forms in Gaussian Random Variables. *IEEE Trans. Commun.*, 64, 1, 153–165, Jan. 2016.

12

Wearable Antennas for Biomedical Applications

Ajay Kumar Singh Yadav[1]*, **Mamta Devi Sharma**[2,3], **Namrata Saxena**[2,4] **and Ritu Sharma**[2]

[1]*Department of ECE, Jaipur Engineering College and Research Center, Jaipur, India*
[2]*Department of ECE, Malaviya National Institute of Technology, Jaipur, India*
[3]*Department of ECE, Global Institute of Technology, Jaipur, India*
[4]*Swami Keshvanand Institute of Technology, Management, Jaipur, India*

Abstract

Since last two decade the research on wearable antennas is carried out more prominently because these antennas are play an important role in 5G communication system for wearable applications, IoT, and biomedical systems due to their unbeatable features like light in weight, manufactured with low-cost flexible substrates, involved simple fabrication techniques and portable. As wearable antennas are work in specific environment, like in proximity of human body therefore the design considerations and requirements of these antennas must be explicit. The materials for antenna design chosen on basis of various aspects for instance these must not be mechanically deformed while bending, stretched and cramping, withstand indifferent weather conditions (like-rain, dust, show, high temperature etc.). The applications of wearable antenna restricted on operating frequency bands. In this context, presents various wearable antennas which are resonated in VHF, UHF and microwave frequency range for off body, on body and in body communication. An outline is given on fabrication process of wearable antennas, based on selected flexible materials with their merits and demerits. The fabrication techniques like Line patterning, Wet-etching, inkjet printing, screen printing and embroidery etc. are commonly employed for such type of antennas. In this chapter, the major requirements of wearable antenna in wireless communication, their applications in biomedical and design criteria, operating frequency bands, fabrication techniques and measurement approaches are emphasized. A detail review on recent development in wearable Textile/Non textile antennas for WBAN are also highlighted.

Corresponding author: ajayrfmicrowave@gmail.com

Arun Kumar, Manoj Gupta, Mahmoud A. Albreem, Dac-Binh Ha and Mohit Kumar Sharma (eds.)
Wearable and Neuronic Antennas for Medical and Wireless Applications, (217–248) © 2022 Scrivener Publishing LLC

Keywords: Wearable antennas, biomedical applications, textile antenna, WBAN, SAR, near body communication, RF spectroscopy, flexible antennas

12.1   Introduction

Since the last two decades, the wearable electronic devices have attracted more attention which offers an incredible opportunity for personal health monitoring, telemedicine, sports, lifestyle and tracking activities. These devices have become an essential function of our daily life and it also supports the 5G communication which is improving the quality of our day-to-day life such as tracking or remote monitoring the biological signals like blood pressure, temperature, heart rate, and respiration rate of a patient. To perform high speed data transport over wireless network, these medical devices involve compact and flexible antennas. Such antennas are known as body-worn antenna, textile antenna, body area-network (BAN) antenna, body centric antenna, and wearable antenna. The first wearable antenna was "planar inverted F" that was manufactured on a flexible dielectric material for dual-band applications in 1999 [1]. The methods involved for design of flexible or wearable antennas are moderately diverse from conventional antennas because these are operated in specific type of environments.

Body Area-Network (BAN) antennas, for personal health care are categorized into three groups [2]:

1) *In-body antennas:* Antenna imbedded in human body for monitoring cardiac activity, brain activity, ingestible or implanted antennas [3].

2) *On-body antennas:* These antennas are momentarily or forever fixed on the external surface of body or are incorporated into wearable garments and various fashionable items.

3) *Off body antennas:* The antenna attached on monitoring/control equipment such as mobile or base station and established a short range communication on link with home/hospital system or distant wireless link amongst hospitals/clinics.

Usually, devices involved in Wireless Body Area Network system collect physiological signal from sensor which are inserted inside the body or attached on body and communicate with external base station. For further processing, these biological signals can be available directly on a user's mobile or desktop and sent over radio link. Figure 12.1 exhibits a Body Centric Communication system with their sub domains.

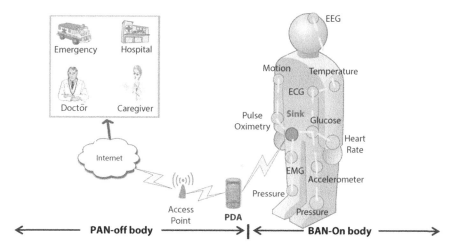

Figure 12.1 Body centric communication [4].

12.2   Need of Wearable Antennas

After the successful achievement of wearable industry with products like smart watches [5], fitness trackers [6], and activity trackers [7], "some emerging products like Owlet's smart sock to monitor infant's heart rate and oxygen levels, footware by Digit sole, anti-back-pain clothing by Percko, smart bikinis, dresses, and swimsuits by Spinali Design and the Iilation jacket" by Teiimo [8], etc. are also available in market as commercial products which are working as smart clothing. Next-generation of wearable wireless communication needs invisible, unobtrusive, flexible and robust sensors which are integrated into garments or accessories. Also the forthcoming procreation of wireless medical devices demand are increasing exponentially for continuous monitoring the real-time psychological signals of patient/individuals and transfer this data on short-distance with high-speed data and low power consumption. The wearable electronic devices may track and monitor the human health without overloading the patient with extra pieces of equipment. The wireless medical equipment must be attached on patient body in an unobtrusive manner, to track the vital parameters of wearer (like-heartbeat, body temperature, respiration activity, etc.), detect, analyze and alert to health advisors/doctors.

The flexible electronic devices include stretchable, thin, light weight low cost antennas that permit numerous bio signals to be precisely measured from human skin and transmitted wirelessly withoutrestricting normal activity. These antennas are the backbone of every wireless wearable device as they manage the stability of the wireless link and openly influence the energy efficiency and power consumption of the system [9].

12.3 Design Considerations for Wearable Antenna

The antenna design for wearable electronics is a significant task because the functioning of an overall wearable system can be modified owing to the human body motion as these structures mounted directly over body or attached on garments. The antenna parameters like gain, radiation pattern impedance can imitated, when antenna operating in existence of human body [10]. Hence, it is

Table 12.1 Wearable antenna design step.

Parameters	Properties
Substrate Material selection	• Easily attached or adhesive with cloths and accessories. • Textile materials should have "small dielectric constant value, minimum surface wave losses and wide impedance bandwidth" • Flexible and Lightweight • High thermal conductivity
Conductive Materials Selection	• High conductivity • Water proof • Corrosion Resist
Antenna Structure/Type	• Simple and compact Structure • Fashionable • Microstrip Patch, Planar Inverted F, Fractal Printed, RFID Loop and Slot antennas etc are mostly used structures for reduce the size and enhance the radiation characteristics
Antenna Performance	• Independent of Human Body Morphology • Stable Under Adverse Conditions "(e.g., Bending, Compression, etc)"
SAR	• Must follow the recommendations given by the FCC and ICNIRP organizations for near-field exposure

a crucial task to recognize the factors that are important to attain the preferred performance of device with stability. Table 12.1 summarizes the key features related to design process for wearable antennas.

The following design considerations are discussed in this section in details for wearable antenna design and fabrication:

(1) *Antenna Frequency Detuning*: Most of the wearable devices working in ISM band at 2.45 and 5.8 GHz for WLAN/5GHz IEEE 802.11 WLAN bands often require single or dual band operations. These narrow bands resonance frequency can be shifted due to loading effect imposes by the human body. These frequency detuning generally occur by interaction among the antenna structure and near-body environment effects. "The human body itself is behaving as lossy and high dielectric medium. Therefore, the designer needs to deliberate the frequency band detuning due to the human body effect". In [11–13], implanted antennas with diverse configurations have been described and antenna performance has been investigated. In [12] a study has been done at thirty different locations in head, trunk and limbs for the detuning and impedance discrepancy of antennas imbedded inside the human body (S_{11}) at different head locations shown in Figure 12.2.

(2) *Radiation Characteristics*: The wearable devices transmit data through WBAN in three circumstances, in-body, on-body, off-body as described in Section 12.1. It is

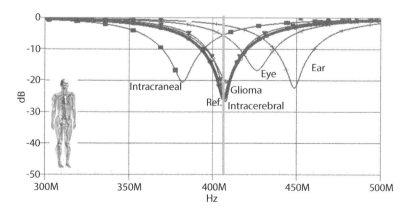

Figure 12.2 S_{11} for the "implanted antenna inside the head of the male anatomical computer model" at the MISC reference frequency [12].

essential to achieve distinct radiation characteristics for different operating mode on the human body, like omnidirectional pattern for on-body configuration and broadside radiation characteristic for in-body and off-body configurations [14]. By adjusting the current flow and introducing a full ground plane, preferred radiation features can be realized that are appropriate for on-body communications.

(3) *Specific Absorption Rate (SAR):* "The SAR is amount of energy absorbed by the human tissue during exposure to a radio frequency (RF) electromagnetic per unit mass" [15]. Exposure of the human body to a radio frequency field results in SAR which is proportional to the squared electric field strength value induced in the body. The higher SAR value results the more absorbed radiations in the tissues and thus the more severe effects on the human body. The SAR is a function of many factors such as frequency and polarization of RF signal, features of the body and features of environment [16]. So the designer must address the consequences of SAR on human body first when design wearable and implantable devices and diminish SAR levels with maintaining the radiation features of the antenna.

(4) *Miniaturization and Cost:* WBAN devices usually require a specific degree of miniaturization to expand the wearer comfort, not restrict mobility and placement location particularly for implanted devices. The size of antenna effects the radiation characteristics, bandwidth ad gain so accurate analysis of size reduction done essentially by designers. "It is a great challenge to model the implantable antennas with the suitable features of biocompatibility, miniaturization, safety, and consistent communications" [17]. Planar inverted F antennas and multi layers PIFA are emerging approach to minimize the volume of in body devices proposed by several researchers in [18, 19]. Other approaches like employment of high dielectric permittivity substrate or superstrate, SIW technology, fractal design, capacitive and inductive loading to improve impedance matching is investigated [20–24]. The other important factor is cost that need to be considered. From a commercial point of view, low cost solution can attract large number of customers.

(5) *Mechanical Robustness:* Furthermore wearable antennas are typically worn by human or living beings, as attached on cloths and accessories or placed inside the body. "If antennas are made-up by conventional lithography or etching metal patterns on rigid substrates, which can easily crease and even fail to function properly when subjected to mechanical deformation (e.g., stretching, folding, or twisting). Thus development of flexible, stretchable, and conformal antennas are required for new electronic materials and/or new device configurations because bending effects degrade the performance of antennas compared to their flat condition" [25]. In [26], the author suggested that when concentrating on a wearable implementation, "it is important to analyze bending effect of the patch antenna when placed at different body locations, such as the leg, arm, or wrist" [27–30].

12.4   Materials for Wearable Antenna

The major problems encountered to design the wearable antennas are size, robustness, less amount of mechanical deformation when bending, twisting and wrapping on human body, resistance when washing and ironing. These can be resolved by applying a suitable selection of materials. To achieve excellent radiation characteristics, efficiency with user comforts and flexibility many scholars have explored various new materials and characterized them. These emerging materials have very specified chemical and electrical properties. There are two kinds of materials highly preferred for fabrication of wearable antennas: (i) conductive as radiating elements and (ii) dielectric material as support platform for radiating part. The materials are broadly categorized into two types such as fabric (E-textile) and non-fabric (like paper, polymers and flexible PCB, etc.) materials.

12.4.1   Fabric Materials

Wireless communication is made possible in the present era because of E-Textile which is treated as dielectric and conductive materials. Fabric Materials are more promising approach for flexible and wearable applications because they can be easily incorporated into clothes, invisible to the user, can be low profile not limiting the wearer's moments and require low cost fabrication. Textile materials should be preferred if wearable antennas are embedded into cloths such as "T-shirts, pants, skirts, and other garments" as similarly nonconductive textile materials are embedded in buttons, glasses, jewelry, and other accessories with metallic parts. "The antenna performance such as impedance bandwidth, radiation efficiency and gain are depended on choice of the dielectric material". Hence, electromagnetic characteristics of the textile dielectric substrates can be characterized first. In the last few years, many investigators have suggested several experimental methods for characterization of conventional fabrics and studied the properties of textiles as dielectric materials [31–40]. "Felt, cotton, jeans fleece, silk, polyester, Cardura, etc. are used as textile substrate materials and to resist from corrosion on a Nickel plated fabric, to achieve strong and flexible property silver plated fabric, Flectron compromise from Copper-coated nylon fabric, and Nora conductive fabrics are used for radiating part of antenna. Table 12.2 exhibits recently developed textile materials for wearable antennas. The achieved results show that conventional fabrics are used as flexible material for textile antennas owing to its low permittivity and a low loss tangent feature [31, 32].

Fully textile antennas are classified into the following categories on basics of materials used for conductive parts:

(1) *Embroidered Textile Antennas:* Theses antennas are fabricated by using digitized embroidery machines, it leads many advantages over other manufacturing techniques as (i) glue is not needed to attach on cloths, (ii) structures are aesthetic and (iii) provide suitable flexibility and

Table 12.2 Recently developed textile materials for wearable antennas.

Ref.	Dielectric material (ε_r)	Conductive material	Frequency band	Applications
[26]	Denim (1.6)	Copper- and nickel-plated polyester	WLAN	WBAN
[33]	Cotton (1.7)	Copper	UWB	Wireless Medical Applications
[34]	Felt Fabric (1.45)	Shieldlt Super	UWB	Off-Body WBAN
[35]	Cardura (2.05)	Taffeta	UWB	Military application

mechanical strength when high tensions are imposed by embroidery machine [36–40].

(2) *E-textiles-based antennas:* "Two techniques are generally used to manufacture E-textile antennas: 1) the weaving process is applied for conductive coating/plating on the surface of a nonconductive textile and 2) embroidery in the textile structure" [41]. The numerous e-textiles are existing in the market with different properties like conductivity, thickness, flexibility, strength and self-adhesive e-textiles also exist. A suitable conductive fabric "must exhibit very low electrical surface resistivity (≤1 Ω per square)".

(3) *Nonwoven Conductive Fabrics (NWCFs) based antennas:* "NWCFs fabrics are long conductive fibers casually arranged and attached together by chemical or mechanical processes". NWCFs demonstrate similar electrical and mechanical properties as e-textiles. NWCFs have numerous advantages: (i) these are exhibiting no fraying problems; (ii) This supports the shaping of self-adhesive foils of nonwoven conductive fabrics with cutting plotter which leads cost-effective and time-saving with large-scale production manufacturing process; (iii) NWCFs materials can be soldered easily so because of this property they allow antennas integrated with sensors and with electronic chips [42].

(4) *Inkjet- and Screen-printed antennas:* In the last few years, inkjet printing and screen printing manufacturing methods are investigated for wearable antennas. By employing these procedures, the antenna layout is directly printed on the textile substrate within few minutes. This will reduce considerable fabrication time and assembly issues [43–45].

12.4.2 Non Fabric Materials

To fabricate non-fabric based antenna one can use soft PCBs for example FR-4, Rogers 3210, Taconic, flexible films like polyimide (PI) films, Polyethylene Terephthalate (PET) films, liquid crystal polymer films (LCP), polymer based substrate like Polydimethylsiloxane (PDMS), paper and foam as dielectric materials whereas conductive nanoparticles such as copper, silver, and gold with polymer deposited/printed as conductive part. In the literature various non-textile antennas are reported as:

(1) *Polymer Based Antenna:* Polymer films have "high flexibility, low loss tangent and available in with low thicknesses". Polymers can be used in the form of silicone elastomers or mixed with ceramic composites to get PDMS substrate. They can also be added to other composites like silver, carbon, and other polymers with different properties. In the last decade, "research on conformal antenna and wearable antennas has become popular by using flexible materials".

A triple band CPW fed antenna for WLAN, Wi MAX and 5th-Generation (5G) systems [46], an UWB triple band notched antenna [47], a series-fed two-dipole antenna [48], an array of microstrip patch antenna at Terahertz frequency [49], and a wearable monopole antenna for WBAN application [50] are fabricated on "Liquid crystal polymer (LCP) dielectric substrate". "LCP is commonly used as a flexible substrate material due to its low dielectric loss, low thermal expansion coefficient, and lower production cost".

Polyethylene Terephthalate (PET) is also widely used in wearable electronics devices. It offers many attractive features such as "highly transparent, robust, conformable, and suitable for printed flexible substrate, etc." A CPW fed antenna for 2.45 GHz frequency, "an integrated far- and near-field communication (FNFC) antenna" in [51] using PET substrate and silver conductive ink for radiating element. A "4 × 4 microstrip patch array antenna" is designed on PET substrate to achieve circularly polarization and high-gain in [52], and "LC wireless humidity sensors fabricated on PET" substrate in [53].

PDMS is emerging material as highly flexible substrate for wearable antenna with offering other advantages on other polymers dielectrics such as water-proof, heat-resistance, UV-rays resistance and chemical stability.

Recently many scholars have demonstrated different types of wearable/flexible antennas by using PDMS as dielectric and transparent conductive thin films, nylon rip stop fabric coated with conductive ink, silver nanowires as conductive materials [54, 55]. Various types of wearable/conformal antennas for single band, dual, multi band and UWB band applications are reported on a Kapton Polyimide substrate which is promising approach to design a compact, flexible, mechanically robust structure with low dielectric losses [56–58]. A variety of flexible antenna based on different fabrication topologies are presented in Figure 12.3.

(2) *Paper Substrate Based Antenna:* In the last few years, paper as flexible substrate which varies in density, coating, thickness, and texture is highly demanded for electronic devices.

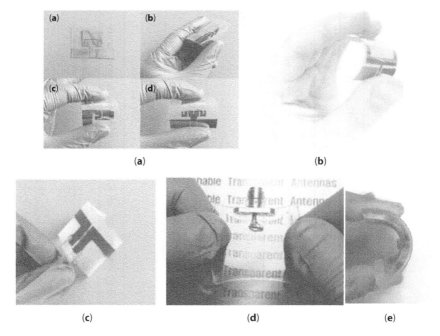

Figure 12.3 Distinct techniques for flexible antennas: (a) PET substrate and silver conductive ink [51]. (b) Ag/NW embedded PDMS [53]. (c) Polymer-Embedded Conductive Fabric [54]. (d) Lycra fabric on porous film [55] (e) Flexible Polyimide substrate [57].

Mass production is possible with organic paper based material because it is widely available at low cost. Fabrication of devices with paper substrate is also fast due printing processes, instead of the conventional metal etching methods. Paper based antenna is also an emerging step of "green" electronic devices due to environmentally friendly nature of paper substrate. Its other advantages such as "low profile, small thickness, and light weight," makes it more appropriate for modern RF devices, such as "Radio-frequency Identification (RFID) tags, wearable antennas, microwave filters and modules". A paper substrate based antenna for wireless sensor networks and RFID systems is designed [59]. Using inkjet-printing process a tri band U slot antenna for "GPS, WiMAX, Hiper LAN/2, and WLAN" is printed on low cost paper substrate in [60].

To design the implantable and flexible antenna a variety of such substrates are listed in Table 12.3. However substrates with high-permittivity (substrate/materials) e.g., ceramic alumina or Rogers 3210 can miniaturized

Table 12.3 Recently developed non textile materials for wearable antennas.

Ref.	Dielectric material (ε_r)	Conductive material	Frequency band	Applications
[46]	LCP (2.9)	Copper	WLAN,WiMax and 5G	Wireless Applications
[51]	PET (3.5)	Silver nano-particle ink	2.45 GHz	Off-Body WBAN
[55]	PDMS (2.8)	Woven conductive fabric	ISM 2.45 GHz	Wearable applications
[57]	Kapton Polyimide (3.5)	Silver nano-particle ink	"GSM 900, GPS, UMTS, WLAN, ISM, Bluetooth, LTE and WiMAX"	Wireless Applications
[60]	Paper (3.2)	Silver ink with nanoparticles hydrocarbon	GPS, WiMAX, HiperLAN/2, and WLAN	Wireless Applications
[62]	RT/Duroid 5880 (2.2)	Copper	UWB	Wireless Medical Applications
[63]	FR-4 (4.4)	Copper	"ISM (2.4 GHz and 5.8 GHz)"	Medical Applications
[64]	Foam Clad (1.02)	Copper foil	WLAN/Wi-Fi	WBAN applications

the antenna structure at lower resonance frequencies. The biocompatible materials includes MACOR and ceramic alumina for providing insulation between conductive part of antenna and human tissues [61].

It is concluded without any doubt that the choice of substrate material is of paramount importance in the realization of flexible antennas. "Due to their conformal behavior and operational suitability, flexible materials have gained immense interest. These flexible materials need to be chosen carefully to withstand the physical deformation conditions such as bending, stretching, and even twisting while maintaining its functionality. Flexible antennas require low-loss dielectric materials as their substrateand highly conductive materials as conductors for efficient EM radiation reception/transmission".

12.5   Fabrication Methods for Wearable Antenna

The fabrication process adopted for flexible and wearable antennas depends on the choice of "conductive and substrate materials" which are discussed in the previous section. These approaches are being used for fabrication of flexible antennas to confirm low cost, robustness and comfort to the wearers in their daily outfit. In this section a brief review of the commercial approaches used by the industries for wearable antennas.

12.5.1   Stitching and Embroidery

The fully textile antennas are fabricated by a computer-aided embroidery machine with aesthetic shapes using colored threads on textile base substrate material. Special type conducting threads are used for radiating element of antennas that can be embroidered onto the base textile. A self-automated embroidery machine is shown in Figure 12.4. The conducting threads must have characteristics like it has flexibility, mechanically elastic, high density of stitching, low resistivity, etc. [65].

Before the embroidered textile antenna fabrication, conductive yarns characterization of the textile substrate is necessary. An UWB antenna has been designed by "embroidered, multifilament thread composed of polyester yarn wrapped by a conductive fiber, which is a composite of copper–nickel alloy that exhibits good conductive and thermal" properties in [66] short-range communication system and a comparative study also have been experimentally performed on embroidered antenna and the traditional printed antenna (built on an FR4 substrate) as shown in Figure 12.5.

Figure 12.4 Embroidery machine [65].

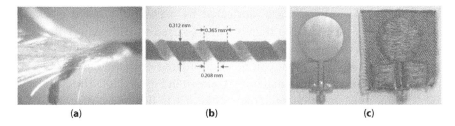

Figure 12.5 Prototype of embroidered UWB antenna with electro conductive polyester yarn: (a) The polyester yarn (b) detail dimensions (c) Fabricated UWB antenna on FR-4 and Cotton [66].

A fully textile UHF RFID on polyester substrate by embroidery [67] and "a dual-band textile antenna" have been embedded for Smart and Sustainable Coat [68]. Figure 12.6 shows the Fabricated "E-Caption: Smart and Sustainable Coat" with measurement results and measurement set up.

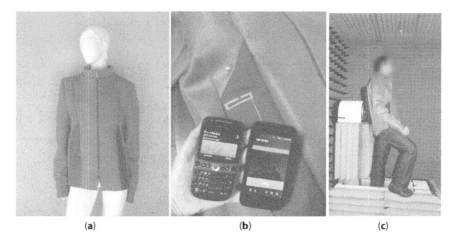

(a)   (b)   (c)

Figure 12.6 "E-Caption: smart and sustainable coat. (a) Design of the coat and (b) integrated antenna, in detail. (c) on-body measurements" [68].

12.5.2   Screen Printing

This technique is also another fast and simple fabrication technique for wearable antennas. In this manufacturing process, the ink is forced through a screen and on the substrate by using a rubber-edge blade, into the exposed areas, to formed pattern [69–74]. The screen printing fabrication techniques has advantages like (i) relatively inexpensive method, (ii) Easy process take less time, (iii) provide good resolution and (iv) mass production possible.

As a summary, "despite being simple, the screen printing technology faces several limitations". "They include its low printing resolution, the limited number of realizable layers and lack of thickness control for the conductive layer". These factors resulted in the limited implementation of such technique, as the printing technology for wearable's requiring "better precision for proper operation of communication front-ends".

12.5.3   Inkjet Printing

In this technique, the antenna pattern on flexible substrate is designed by conductive Nanoparticle ink droplets using a smart printer which integrated with nozzle of Pico range diameter. The deposited ink can be thermally cured up to 150 °C to confirm the pattern for antenna application. The process flow of this technique is summarized in Figure 12.7 [75]. This fabrication process

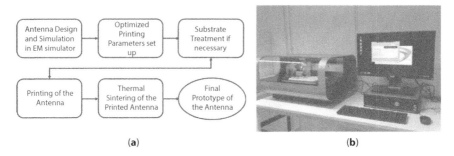

(a)   (b)

Figure 12.7 (a) Fabrication flow chart (b) specialized DMP 2831 printer.

is very costly and it is a major challenge to maintain pattern conductivity stable after the curing temperature. The printing is performed using a specialized "DMP 2831 printer" as shown in Figure 12.7.

12.5.4   Chemical Etching

Photolithography is used in chemical etching process for away the unnecessary area to construct the desired metallic patterns (Figure 12.8 (a)). This technique has widely acceptable in manufacturing industry owing to it can generates patterns with high resolution and fine pattern details. With the development of the "Printed Circuit Board (PCB) industry chemical etching" came into market in the1960s.

The chemical photoresists are utilized for the etching process, its chemical characteristics are change on exposure of ultraviolet light like organic polymers. There are two types of photoresists utilize like positive photoresist and negative photoresist, subsequently the goal of negative photoresist to start the swelling occurrences, which negotiates the resolution of the produced pattern. Although in flexible electronics, the chemical etching method is extensively used to produce flexible antennas or devices; this technique requires hazardous chemicals but it creates negative effects to environment [76–78].

In [77] a circularly polarized microstrip UWB antenna has proposed for wearable applications as conformal structure. Figures 12.8 (b and c) shows measured reflection coefficients of the fabricated antenna and compare this result with simulated values. Simulated and on body results are also presented in same graph as seen in Figure 12.8 (c).

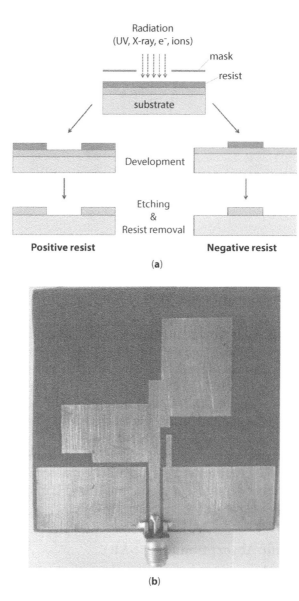

(a)

(b)

Figure 12.8 (a) Fabrication process (b) UWB antenna printed on substrate by etching and (c) S_{11} Vs frequency graph [77].   (*Continued*)

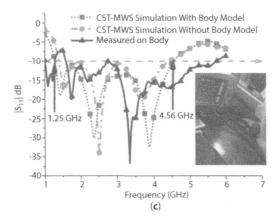

Figure 12.8 (Continued) (a) fabrication process (b) UWB antenna printed on substrate by etching and (c) S_{11} Vs frequency graph [77].

12.6 Measurements for Wearable Antenna

To validate the performance characteristics of the conventional designed structure basic parameters like "S-parameters (reflection coefficient and transmission coefficient), radiation patterns, efficiency, polarization and gain, etc." are measured after fabrication with help of Vector Network Analyzer (VNA) and anechoic chamber. Apart from this for wearable antennas, other parameters also evaluate and test such as SAR, robustness of design, durability, bending, and crumpling effects in different environment and with human proximity as they are generally needed by the "Federal Communication Commission (FCC)" and quality control department. Some important parameters are discussed below.

12.6.1 Specific Absorption Rate (SAR)

"For wearable applications, the effect of antennas on human body must be quantified". SAR is defined as the maximum EM power absorption by the human body tissues during RF exposure from wireless radiation. The averaged SAR measures in W/kg or mW/g terms, over a specific volume of 1 or 10 g of tissue. A whole human body phantom or two small cubes as phantom is employed for *vitro* measurement. A detail acquisition technique exploit in "SAR measurement" is described in [79]. Here a phantom filled with a synthesized liquid which has electrical properties as mimic the human body. The FCC limits the SAR values at different bands for human safety. According to FCC, if SAR value <1.6 W/kg over 1 g sample is consider as safe limit of exposure (in the United States), while in Europe, safe level of SAR is regulated at 2.0 W/kg or less over 10 g tissue sample. Various researchers have chosen different phantoms (e.g. rectangular or cylindrical types cubes of 1 or 10 g tissues with different electrical characteristics and different sized phantoms), tissue models with different tissue compositions, human models (with different heights and weights) and kept antenna at different separations from the phantom/human model, to analysis of SAR values. Recently, two extensive ways are utilize for SAR analysis: (i) Electric-field probe method and (ii) thermographic method [80, 81].

12.6.2   Performance on Human Body

"Measured results of wearable antennas and radio frequency (RF) medical systems in the proximity of the human body are represented" in [82, 83]. Antenna parameters like: operating frequency, reflection losses, radiation efficiency, and gain are mostly affected by the proximity of human tissues. For real time applications, the simulated antenna should be tested in the human body environment and evaluate coupling effects between antenna structure with human body. In [82], two UWB homogeneous semisolid phantoms "(namely, a flat phantom with dimensions of 200 mm × 200 mm × 47 mm and a human-hand-shaped phantom)" as seen in Figure 12.9 were employed to measure the antenna performance near human proximity.

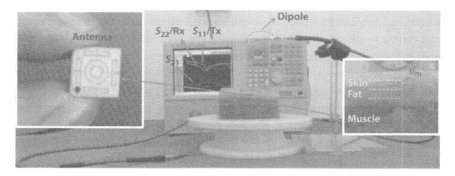

Figure 12. 9 Photograph of the fabricated antenna prototype and of the measurement setup [83].

12.6.3   The Bending and Crumpling Effects

The most inevitable challenge to design flexible and wearable antennas for WBAN applications are mechanical deformation that occurs owing to bending, crumpling and sometimes, twisting actions the structure on human body or with movement of body. For practical applications, the antenna output parameters alter, often negatively; "this can be explained by the electric field distribution. Hence, some assessments are necessary to be evaluated for working reliability of system". These are reported with a detailed procedure for flexible antenna tests in [84, 85] and can be summarized as follows:

> ➢ Durability and robustness tests
> ➢ Return loss tests in bending conditions
> ➢ Patterns, gain and directivity of the antenna.

"The effect of crumpling in different directions is inconsistent. Under bending and crumpling case, the operating frequency detuned as well as radiation efficiency varies and radiation patterns are tilted and/or distorted". The extension in bending and crumpling produces the huge amount of distortion. The reader is referred to [86, 87] for additional useful discussions in this area.

12.7   Frequency Bands for Wearable Antenna

In Table 12.4, the wearable antennas topologies with their feasible frequency bands and possible applications are summarized.

"Wireless Medical Telemetry System (WMTS) and Medical Implant Communications Service (MICS) bands were allocated entirely for body-worn and implanted medical applications, which involve simple point-to-point communication". These bands are provided high data rate and consistent communication and overcome magnetic coupling interference problem. These are considering as the early wireless medical devices. The MICS band is more suitable for short rage telemetry communication applications with maximum data rate up to 400 kbps such as pacemakers for heart rate measurement, implanted defibrillator and neuro stimulator. Biomedical telemetry applications like the swallowable camera pill operated in the WMTS band with 1 Mbps data rate.

The International Telecommunication Union (ITU)is defined industrial, scientific and medical (ISM) bands (2.4–2.5 GHz and 5.72–5.87 GHz)

Table 12.4 "Frequency bands used by antennas for in-body, on-body and external medical device connectivity" [88–92].

Type	Frequency band	Data rate	Applications	Range (m)
In body (Implanted Antenna) UHF/VHF MICS WMTS	13.56 MHz, 840–960 MHz, 401-402,402-405, 403.5-403.8(MITS) and 405-406 (MEDS) MHz "608-614 MHz, 1395-1400 MHz and 1427-1432 MHz"	Very low	Implant biotelemetrypace maker to monitor abnormal heart beat, capsule endoscopy, for diagnosis and/or therapy	<=0.5
Off body ISM band WiMax UWBband	2.4–2.48 GHz,5.725 – 5.875 GHz, 2.3/2.5/3.5 GHz 3.1 -10.6 GHz	Very High	"Indoor navigation for patients, connectivity btw device and smartphone for health data monitoring, Hospital intranet, Telemedicine services btw hospitals/clinics or ambulances/ hospitals for pre-diagnosis"	Greater then 1m and less then 10 m
On body ISM band UWBband MBAN	2.4-2.48 GHz, 5.725 –5.875 GHz, 3.1 -10.6 GHz, 2.36-2.4 GHz	Very High	"Short range biotelemetry implantantennas with dual-frequency operation, cancer detection, neural monitoring, wireless channel characterization in hospital environments"	Greater then 1m

as unlicensed bands for serve WBAN applications for off body and on body communication. "An interesting action was taken by the Federal Communications Commission (FCC) on May 2012 to allocate 40 MHz of spectrum between 2.36 and 2.40 GHz on a secondary basis for a new Medical Body Area Network (MBAN) as licensed services". The other best approach is ultra wideband spectrum for WBAN. According ITU, an UWB spectrum is that occupies more than 500 MHz of spectrum. "The regulatory authority specifies, however, that the power spectral density shall not exceed—41.25 dBm/MHz, which around 30 dB below the maximum is allowed for a signal in the 2.4–2.5 GHz ISM band".

12.8 Applications of Wearable Antenna in Biomedical

In present era biomedical engineering is the fastest developing fields of technology which a field of recent achievements and even more ambitious hopes. Recently, several types of antennas have been studied and reported for medical applications (such as in diagnosis, microwave imaging for cancer detection and treatment of various chronic diseases). Like Implanted sensors can sense any type of abnormal conditions in body and this collected data to be sent on gateway such as a mobile phone or PC (Base Unit). The base unit then transmits this real time data through "cellular network or the internet to a remote place such as an emergency center or a hospital and receives" feedback from doctor or healthcare personal. Moreover, wireless medical devices will be providing best solution for "early detection, diagnosis of disease, monitoring and treatment of real time data of patients, including diabetes, hypertension and cardiovascular related diseases".

Medical Applications of Antennas
Wearable antennas are divided mainly in two categories on basis of biomedical application as

    (i)   Wearable devices for Diagnosis—use for "Biomedical Telemetry propose and for in body sensors to detect or monitor biological objects, providing physiological, motion-related, and location-related information".

    (ii)   Treatment (thermal therapies)—Microwave hyperthermia of cancer patients, Microwave coagulation therapy.

However wearable antenna has a wide range of applications in biomedical industry and it demonstrated by researchers [93–104] and industries. Here a list of such applications is summarized.

> *Biomedical Telemetry Applications*
> *Aiding Professional and Amateur Sport Training*
> *Sleep Staging:* "Sleep is an important behavior and regular physiological function which consumes one-third of our everyday life. A large population have sleep disorders and obstructive sleep apnea (OSA), making it one of the most common sleep disorders. OSA sufferers present severe snoring and poor sleep quality. It also causes daytime sleepiness and chronic fatigue which, in turn, increases the risk of accidents caused by drowsiness. Besides, OSA is closely related to diabetes and contributes to high blood pressure and arterial the associate hardening. Moreover, patients with sleep apnea have a higher risk of degenerative brain diseases, such as heart disease and Alzheimer's. Therefore, sleep monitoring has gained great interest in the recent years. Sleep disorders can be realized through a polysomnography test which requires analysis of a number of biopotentials recorded overnight in a sleep laboratory" [96–100].
> *Monitoring the Athlete activities:* "Advancements in wearable devices and sensors enable athletes, coaches, and physicians to track functional movements, workload, biomechanical and bio-vital markers to maximize performance and minimize the potential for injury".
> *Measurement of human body Temperature:* The temperature of a healthy person ranges between 35 and 38 °C. Temperatures below or above this range may indicate that the person is sick. Temperatures above 40 °C may cause death. A person's body temperature may be transmitted to a medical center and if needed the doctor may contact the patient for further assistance [101–104]. Figure 12.10 presented the possible locations for antenna implantation over the human body for various physical and environmental parameters sensing applications.
> *Measurement of blood pressure:* A blood pressure measurement indicates the arterial pressure of the blood circulating in the human body. Some of the causes of changes in blood pressure may be stress and being overweight. The blood pressure of a healthy person is around 80 by 120, where the systole is 120 and the diastole is 80.
> *Measurement of heart rate:* Measurement of the heart rate is one of the most important tests when examining

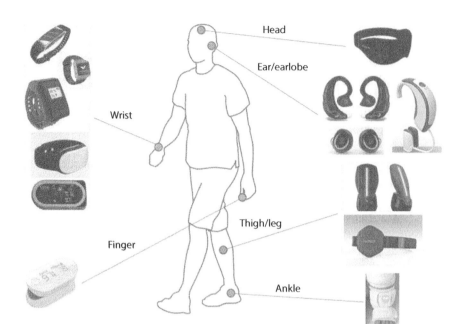

Figure 12.10 Various wearable devices placed on different locations on human body [101].

the health of a patient. A change in heart rate will change the blood pressure and the amount of blood delivered to all parts of the body. The heart rate of a healthy person in 72 beats per minute. Changes in heartbeat may cause several kinds of cardiovascular disease. Traditionally heart rate is measured using a stethoscope. However, this is a manual test and is not so accurate. To measure and analyze the heartbeat a wearable medical device may be connected to a patient's chest. Medical devices that measure heartbeat can be wired or wireless [104–108].

➢ *Wearable devices for respiratory assessment*
➢ *Pressure sensors*
➢ *Acoustic sensors*
➢ *Oximetry sensors*
➢ *Glucose monitoring*
➢ *Cancer Detection*
➢ *Remote Control of Medical Devices*
➢ *Patient Monitoring*
➢ *Telemedicine Systems*
➢ *Wearable Devices for Treatment*
➢ *Hyperthermia therapy.*

12.9   Conclusion

A brief review on the wearable antennas for biomedical and other smart devices application has discussed. To design such antennas it is very auspicious to consider the substrate materials, conductive materials, fabrication process and measurement of these antennas, this chapter included all the necessary steps to implement the antenna for appropriate application. The fundamental characteristics and limitations of wearable antenna along with their solution discussed which give an overview for the designers to consider before a new advancement in biomedical and wearable electronics devices.

References

1. Salonen, P., Sydanheimo, L., Keskilammi, M., Kivikoski, M., A small planar inverted-F antenna for wearable applications. *International Symposium on Wearable Computers*, 1999.
2. Hall, P. and Hao, Y., *Antennas and Propagation for Body-centric wireless Communication*, 2nd Edition, Artech House, Boston, London, 2012.
3. Kiourti, A. and Nikita, K.S., A review of implantable patch antennas for biomedical telemetry: challenges and solutions. *IEEE Antennas Propag. Mag.*, 54, 3, 210–25, 2012.
4. https://images.app.goo.gl/9s8Tnco7AedjWYsa7
5. Reeder, B. and David, A., Health at hand: A systematic review of smart watch uses for health and wellness. *J. Biomed. Inform.*, 63, 296–276, Oct. 2016.
6. Ridgers, N.D., McNarry, M.A., Mackintosh, K.A., Feasibility and effectiveness of using wearable activity trackers in youth: A systematic review. *JMIR Mhealth Uhealth*, 4, 4, e129, Oct.–Dec. 2016.
7. Tedesco, S., Barton, J., OFlynn, B., A review of activity trackers for senior citizens: Research perspectives, commercial landscape and the role of the insurance industry. *Sensors*, 17, 6, 1277, 2017.
8. Lemey, S., Agneessens, S., Rogier, H., Wearable smart objects: Microwaves propelling smart textiles: A review of holistic designs for wireless textile nodes. *IEEE Microw. Mag.*, 19, 6, 83–100, Sep./Oct. 2018.
9. Atanasov, N.T., Atanasova, G.L., Atanasov, B.N., Al-Rizzo, H., Wearable Textile Antennas with High Body-Area Isolation: Design, Fabrication, and Characterization Aspects, in: *Modern Printed Circuit Antennas*, 1st ed, vol. 1, pp. 1–20, IntechOpen: London, UK, 2020.
10. Sabban, A., *Novel Wearable Antennas for Communication and Medical Systems*, CRC Press, ISBN: 9780367889340, December 11, 2019.
11. Furse, C.M., Today's opportunities and challanges. *IEEE Workshop on Antenna Technology Small Antennas and Novel Metamaterials*, Santa Monica, California, March, 2009.

12. Vidal, N., Curto, S., Lopez-Villegas, J.M., Sieiro, J., Ramos, F.M., Detuning Study of Implantable Antennas Inside the Human Body. *Prog. Electromagn. Res. B.*, 124, 265–283, 2012.

13. Feng, Y., Li, Y., Li, L., Ma, B., Hao, H., Li, L., Tissue-Dependent Co-Matching Method for Dual-Mode Antenna in Implantable Neurostimulators. *IEEE Trans. Antennas Propag.*, 67, 8, 5253–5264, Aug. 2019.

14. Mendes, C. and Peixeiro, C., On-body transmission performance of a novel dual-mode wearable microstrip antenna. *IEEE Trans. Antennas Propag*, 66, 9, 4872–4877, 2018.

15. Hamed, T. and Maqsood, M., SAR Calculation & Temperature Response of Human Body Exposure to Electromagnetic Radiations at 28, 40 and 60 GHz mmWave Frequencies. *Prog. Electromagn. Res. B*, 73, 47–59, 2018.

16. International Commission on Non-Ionizing Radiation Protection (ICNIRP)," Guidelines for limiting exposure to time-varying electric, magnetic and electromagnetic fields (up to 300 GHz). *Health Phys.*, 74, 4, 494–522, 1998.

17. Soontornpipit, P., Furse, C.M., Chung, Y.C., Design of implantable microstrip antenna for communication with medical implants. *IEEE Trans. Microw. Theory Tech.*, 52, 1944–1951, Aug. 2004.

18. Lin, C.-H., Saito, K., Takahashi, M., Ito, K., A Compact Planar Inverted-F Antenna for 2.45 GHz On-Body Communications. *IEEE Trans. Antennas Propag.*, 60, 9, 4422–4426, 2012.

19. Bhattacharjee, S., Midya, M., Mitra, M., BhadraChaudhuri, S., Dual band-dual polarized planar inverted F-antenna for MBAN applications. *Int. J. Microw. Wirel. Technol.*, 11, 1, 76–86, 2019.

20. Badhan, K. and Singh, J., Analysis of Different Performance Parameters of Bodywearable Antenna-A Review. *Adv. Wirel. Mobil. Commun.*, 10, 5, 735–745, 2017. ISSN 0973- 6972.

21. Agneessens, S., Lemey, S., Vervust, T., Rogier, H., Wearable, Small, and Robust: The Circular Quarter-Mode Textile Antenna. *IEEE Antennas Wirel. Propag. Lett.*, 14, 1482–1485, 2015.

22. Poonkuzhali, R., Alex, Z.C., Balakrishnan, T.N., Miniaturized Wearable Fractal Antenna for Military Applications at VHF Band. *Prog. Electromagn. Res. C*, 62, 179–190, 2016.

23. Juan, Y., Yang, W., Che, W., Miniaturized Low-Profile Circularly Polarized Metasurface Antenna Using Capacitive Loading. *IEEE Trans. Antennas Propag.*, 67, 5, 3527–3532, May 2019.

24. Motevasselian, A. and Whittow, W.G., Patch size reduction of rectangular microstrip antennas by means of a cuboid ridge. *IET Microw. Antennas Propag.*, 9, 15, 1727–1732, 2015L.

25. Song and Y. Rahmat-Samii, A Systematic Investigation of Rectangular Patch Antenna Bending Effects for Wearable Applications. *IEEE Trans. Antennas Propag.*, 66, 5, 2219–2228, May 2018.

26. Ferreira, D., Pires, P., Rodrigues, R., Caldeirinha, R.F.S., Wearable Textile Antennas: Examining the effect of bending on their performance. *IEEE Antennas Propag. Mag.*, 59, 3, 54–59, June 2017.

27. Gharbi, M., Martinez-Estrada, M., Fernández-García, R., Ahyoud, S., Gil, I., A novel ultra-wide band wearable antenna under different bending conditions for electronic-textile applications. *J. Text. Inst.*, 112, 437–443, 2021.

28. Toh, W.Y., Tan, Y.K., Koh, W.S., Siek, L., Autonomous wearable sensor nodes with flexible energy harvesting. *IEEE Sensors J.*, 14, 7, 2299–2306, July 2014.

29. Chi, Y.-J., Lin, C.-H., Chiu, C.-W., Design and modeling of a wearable textile rectenna array implemented on Cordura fabric for batteryless applications. *J. Electromagn. Waves Appl.*, 34, 13, 1782-1796, 2020, doi: 10.1080/09205071.2020.1787869.

30. Adami, S.-E., Proynov, P., Hilton, G.S. *et al.*, A flexible 2.45-GHz power harvesting wristband with net system output from –24.3 dBm of RF power. *IEEE Trans. Microw Theory*, 66, 1, 380–395, 2017.

31. Paracha, K.N., Rahim, S.K.A., Soh, P.J., Khalily, M., Wearable Antennas: A Review of Materials, Structures, and Innovative Features for Autonomous Communication and Sensing. *IEEE Access*, 7, 56694–56712, 2019, doi: 10.1109/ACCESS.2019.2909146.

32. Corchia, L., Monti, G., Tarricone, L., Wearable antennas: Nontextile versus fully textile solutions. , in: *IEEE Antennas and Propagation Magazine,* vol. 61, no. 2, pp. 71–83, April 2019.

33. Singh, N., Singh, A.K., Kumar Singh, V., Design and Performance of Wearable Ultrawide Band Textile Antenna For Medical Applications. *Microw. Opt. Technol. Lett.*, 57, 7, 1553–57, July 2015.

34. Samal, P.B., Soh, P.J., Vandenbosch, G.A.E., UWB All-Textile Antenna With Full Ground Plane for Off-Body WBAN Communications. *IEEE Trans. Antennas Propag.*, 62, 1, 102–108, Jan. 2014, doi: 10.1109/TAP.2013.2287526.

35. Indumathi, G. and Bhavithra, J., Wearable textile antenna for indoor applications. *2017 International Conference on Inventive Communication and Computational Technologies (ICICCT)*, Coimbatore, pp. 30–34, 2017, doi: 10.1109/ICICCT.2017.7975221.

36. Tsolis, A., Whittow, W.G., Alexandridis, A.A., Vardaxoglou, J.C., Embroidery and related manufacturing techniques for wearable antennas: Challenges and opportunities. *Int. J. Electron.*, 3, 2, 314–338, 2014.

37. Ouyang, Y. and Chappell, W.J., High frequency properties of electrotextiles for wearable antenna applications. *IEEE Trans. Antennas Propag.*, 56, 2, 381–389, 2008.

38. Zhang, S. *et al.*, Embroidered wearable antennas using conductive threads with different stitch spacings, in: *Proc. Loughborough Antennas & Propagation Conf. (LAPC)*, Loughborough, U.K., pp. 1–4, 2012, doi: 10.1109/LAPC.2012.6403059.

39. Toivonen, M., Björninen, T., Sydänheimo, L., Ukkonen, L., RahmatSamii, Y., Impact of moisture and washing on the performance of embroidered UHF RFID tags. *IEEE Antennas Wirel. Propag. Lett.*, 12, 1590– 1593, Nov. 2013.

40. Fu, Y.Y. *et al.*, Experimental study on the washing durability of electrotextile UHF RFID tags. *IEEE Antennas Wirel. Propag. Lett.*, 14, 466–469, Nov. 2015.

41. Seager, R. *et al.*, Effect of the fabrication parameters on the performance of embroidered antennas. *IET Microw. Antennas Propag.*, 7, 14, 1174–1181, 2013.

42. Liu, N., Lu, Y., Qiu, S., Li, P., Electromagnetic properties of electro-textiles for wearable antennas applications. *Front. Elect. Electron. Eng. China*, 6, 4, 563–566, 2011.

43. Whittow, W.G. *et al.*, Inkjet-printed microstrip patch antennas realized on textile for wearable applications. *IEEE Antennas Wirel. Propag. Lett.*, 13, 71–74, 2014.

44. Roshni, S.B., Jayakrishnan, M.P., Mohanan, P., Surendran, K.P., Design and fabrication of an E-shaped wearable textile antenna on PVB-coated hydrophobic polyester fabric. *Smart Mater. Struct.*, 26, 105011, 8pp, 2017.

45. Klemm, M. and Troester, G., Textile UWB Antennas for Wireless Body Area Networks. *IEEE Trans. Antennas Propag.*, 54, 11, 3192–3197, Nov. 2006, doi: 10.1109/TAP.2006.883978.

46. Du, C., Li, X., Zhong, S., Compact Liquid Crystal Polymer Based Tri-Band Flexible Antenna for WLAN/WiMAX/5G Applications., *IEEE Access.* doi: 10.1109/ACCESS.2019.2941212.

47. Xiao, W., Mei, T., Lan, Y., Wu, Y., Xu, R., Xu, Y., Triple bandnotched UWB monopole antenna on ultra-thin liquid crystal polymer based on ESCSRR. *Electron. Lett.*, 53, 2, 57–58, 19 1 2017.

48. Kao, H. *et al.*, Bending Effect of an Inkjet-Printed Series-Fed TwoDipole Antenna on a Liquid Crystal Polymer Substrate. *IEEE Antennas Wirel. Propag. Lett.*, 13, 1172–1175, 2014.

49. Rabbani, M.S. and Ghafouri-Shiraz, H., Liquid Crystalline Polymer Substrate-Based THz Microstrip Antenna Arrays for Medical Applications. *IEEE Antennas Wireless Propag. Lett.*, 16, 1533–1536, 2017.

49. Zahran, S.R., M. A. Abdalla and A. Gaafar, "New thin wide-band bracelet-like antenna with low SAR for on-arm WBAN applications. *IET Microw., Antennas Propag.*, 13, 8, 12191225, 3 7 2019.

50. Xiaohui, G. *et al.*, Flexible and wearable 2.45 GHz CPW-fed antenna using inkjet-printing of silver nanoparticles on pet substrate. *Microw Opt Technol. Lett.*, 59, 1, 204–208, 2017.

51. Castro, A.T. and Sharma, S.K., Inkjet-Printed Wideband Circularly Polarized Microstrip Patch Array Antenna on a PET Film Flexible Substrate Material. *IEEE Antennas Wirel. Propag. Lett.*, 17, 1, 176–179, Jan. 2018, doi: 10.1109/LAWP.2017.2779440.

52. Deng, W., Wang, L., Dong, L., Huang, Q., Experimental Study of the Bending Effect on LC Wireless HumiditySensors Fabricated on Flexible PET Substrates. *J. Microelectromech. Syst.*, 27, 5, 761–763, Oct. 2018, doi: 10.1109/JMEMS.2018.2856912.

53. Kim, B.S., Shin, K.-Y., Pyo, J.B., Lee, J., Son, J.G., Lee, S.-S., Park, J.H., Reversibly Stretchable, Optically Transparent Radio-Frequency Antennas Based on Wavy Ag Nanowire Networks. *ACS Appl. Mater. Interfaces*, 8, 2582–2590, 2016.

54. Simorangkir, R.B.V.B., Yang, Y., Esselle, K.P., Zeb, B.A., A Method to Realize Robust Flexible Electronically Tunable Antennas Using Polymer-Embedded Conductive Fabric. *IEEE Trans. Antennas Propag.*, 66, 1, 50–58, Jan. 2018, doi: 10.1109/TAP.2017.2772036.

55. Guo, X., Huang, Y., Wu, C., Mao, L., Wang, Y., Xie, Z., Liu, C., Zhang, Y., Flexible and reversibly deformable radio-frequency antenna based on stretchable SWCNTs/PANI/Lycra conductive fabric. *Smart Mater. Struct.*, 26, 105036, 2017.

56. Khaleel, H.R., Al-Rizzo, H.M., Abbosh, A., II, Design fabrication and testing of flexible antennas, in: *Advancement in Microstrip Antennas With Recent Applications*, Austria:InTech, Vienna, 2013.

57. Ahmed, S., Tahir, F.A., Shamim, A., Cheema, H.M., A Compact Kapton-Based Inkjet-Printed Multiband Antenna for Flexible Wireless Devices. *IEEE Antennas Wirel. Propag. Lett.*, 14, 1802–1805, 2015, doi: 10.1109/LAWP.2015.2424681.

58. Khaleel, H.R., Al-Rizzo, H.M., Rucker, D.G., Mohan, S.A., Compact Polyimide-Based UWB Antenna for Flexible Electronics. *IEEE Antennas Wirel. Propag. Lett.*, 11, 564–567, 2012.

59. KizhekkePakkathillam, J. and Kanagasabai, M., Performance evaluation of a dual band paper substrate wireless sensor networks antenna over curvilinear surfaces. *IET Microw., Antennas Propag.*, 9, 8, 715–722, 5 6 2015, doi: 10.1049/iet-map.2014.0691.

60. Abutarboush, H.F. and Shamim, A., Paper-Based Inkjet-Printed Tri-Band U-Slot Monopole Antenna for Wireless Applications. *IEEE Antennas Wirel. Propag. Lett.*, 11, 1234–1237, 2012, doi: 10.1109/LAWP.2012.2223751.

61. Kiourti, A. and Nikita, K.S., A Review of Implantable Patch Antennas for Biomedical Telemetry: Challenges and Solutions [Wireless Corner]. *IEEE Antennas Propag. Mag.*, 54, 3, 210–228, June 2012, doi: 10.1109/MAP.2012.6293992.

62. Nadh, B.P., Madhav, B.T.P., Kumar, S. *et al.*, Circular ring structured ultra-wideband antenna for wearable applications. *Int. J. RF Microw. Comput.-Aided Eng.*, 28, 9, 1–15, 2018.

63. Kissi, C. *et al.*, Dual Band CPW-Fed Double Monopole Antenna for 2.4/5.8 GHz ISM band Medical Applications. *2019 International Symposium on Advanced Electrical and Communication Technologies (ISAECT)*, pp. 1–6, 2019, doi: 10.1109/ISAECT47714.2019.9069690.

64. Singh, R.K. and Gupta, A., Design and Development of U-shaped slot Wearable Antenna for WLAN/Wi-Fi and WBAN Applications. *2018 International Conference on Advances in Computing, Communication Control and Networking (ICACCCN)*, Greater Noida (UP, India, pp. 1063–1067, 2018, doi: 10.1109/ICACCCN.2018.8748834.

65. Tsolis, A., Whittow, W., Alexandridis, A., Vardaxoglou, J., Embroidery and Related Manufacturing Techniques for Wearable Antennas: Challenges and Opportunities. *Electronics*, 3, 314–338, 2014.

66. Sim, C.Y.D., Tseng, C.W., Leu, H.J., Embroidered wearable antenna for ultrawideband applications. *Microw Opt Technol. Lett.*, 54, 2597–600, 2012.

67. Abbas, B., Khamas, K., Ismail, S., Sali, A., Full embroidery designed electro-textile wearable tag antenna for WBAN application. *Sensors*, 19, 2470, 2019.

68. Loss, C., Gonçalves, R., Lopes, C., Pinho, P., Salvado, R., Smart Coat with a Fully-Embedded Textile Antenna for IoT Applications. *Sensors*, 16, 938, 2016.

69. Merilampi, S.L., Virkki, J., Ukkonen, L., Sydänheimo, L., Testing the effects oftemperature and humidity on printed passive UHF RFID tags on paper substrate. *Int. J. Electron.*, 101, 5, 711–730, 2014. http://dx.doi.org/10.1080/00207217.2013.794491.

70. Shin, D.-Y., Lee, Y., Kim, C.H., Performance characterization of screen printed radio frequency identification antennas with silver nanopaste. *Thin Solid Films*, 517, 6112–6118, Sep. 2009.

71. Virkki, J., Björninen, T., Kellomäki, T., Merilampi, S., Shafiq, I., Ukkonen, L., Sydänheimo, L., Chan, Y.C., Reliability of washable wearable screen printed UHF RFID tags. *Microelectronic. Reliab.*, 54, 840, 2014.

72. Roshni, S.B., Jayakrishnan, M.P., Mohanan, P., Surendran, K.P., Design and fabrication of an E-shaped wearable textile antenna on PVBcoated hydro-phobic polyester fabric. *Smart Mater. Struct.*, 26, 10, 105011, 2017.

73. Xu, D., Tian, X., Guo, X., Jiang, W., Liu, W., Xing, S., Design and research of flexible wearable textile antenna based on GNPs/PANI/PDMS composites for 2.45 GHz. *Nanosci. Nanotechnol. Lett.*, 9, 4, 476–480, 2017.

74. Shahariar, H., Soewardiman, H., Muchler, C.A., Adams, J.J., Jur, J.S., Porous textile antenna designs for improved wearability. *Smart Mater. Struct.*, 27, 4, Art. no. 045008, 2018.

75. Kirtania, S.G., Elger, A.W., Hasan, M.R., Wisniewska, A., Sekhar, K., Karacolak, T., Sekhar, P.K., Flexible Antennas: A Review. *Micromachines.* 11, 847, 2020.

76. Khaleel, H., Abbosh, A., Al-Rizzo, H., ch. 15, in: *Design, Fabrication, and Testing of Flexible Antennas.*

77. Das, S., Islam, H., Bose, T. *et al.*, Ultra Wide Band CPW-Fed Circularly Polarized Microstrip Antenna for Wearable Applications. *Wirel. Pers. Commun.*, 108, 87–106, 2019. https://doi.org/10.1007/s11277-019-06389-9.

78. Board, N., *Handbook On Printing Technology (Offset, Gravure, Flexo, Screen)*, 2nd edition, Asia Pacific Business Press Inc, New Delhi, 2011.

79. Zhang, Z., *Antenna Design for Mobile Devices*, pp. 239–259, John Wiley & Sons, Singapore Pte. Ltd, 2011.

80. Merckel., O., Gilles. Fleury, G., Bolomey, J.-C., Propagation model choice for rapid SAR measurement. *Signal Processing Conference*, pp. 1–4, 2002.

81. Guy, A.W., Analysis of electromagnetic fields induced in biological tissues by thermographic studies cmequivalent phantom models. *IEEE Trans. Microw. Theory Tech.*, MTT-19, 205–214, Feb. 1971.

82. Sayem, A.S.M., Simorangkir, R.B.V.B., Esselle, K.P., Hashmi, R.M., Development of Robust Transparent Conformal Antennas Based on Conductive MeshPolymer Composite for Unobtrusive Wearable Applications. *IEEE Trans. Antennas Propag.*, 67, 12, 7216–7224, Jul. 2019.

83. Liu, X.Y., Wu, Z.T., Fan, Y., Tentzeris, E.M., A Miniaturized CSRR Loaded Wide-Beamwidth Circularly Polarized Implantable Antenna for Subcutaneous Real-Time Glucose Monitoring. *IEEE Antennas Wirel. Propag. Lett.*, 16, 577–580, 2017, doi: 10.1109/LAWP.2016.2590477.

84. Roh, J., Chi, Y., Lee, J., Tak, Y., Nam, S., Kang, T.J., Embroidered Wearable Multi resonant Folded Dipole Antenna for FM Reception. *IEEE Antennas Wirel Propag. Lett.*, 9, 803–806, 2010, doi: 10.1109/LAWP.2010.2064281.

85. Khaleel, H.R., Al-Rizzo, H., Rucker, D., Compact polyimide based antennas for flexible displays. *IEEE J. Display Technol.*, 8, 2, 91–97, 2012.

86. Bai, Q. and Langley, R., Crumpling of PIFA textile antenna. *IEEE Trans. Antennas Propag.*, 60, 1, 63–70, 2012.

87. Bai, Q., Rigelsford, J., Langley, R., Crumpling of microstrip antenna array. *IEEE Trans. Antennas Propag.*, 61, 9, 4567–4576, 2013.

88. Cavallari, R., Martelli, F., Rosini, R., Buratti, C., Verdone, R., A Survey on Wireless Body Area Networks: Technologies and Design Challenges. *IEEE Commun. Surv. Tutor.*, 16, 3, 1635–1657, Third Quarter 2014, doi: 10.1109/SURV.2014.012214.00007.

89. The Order on Reconsideration and Second Report and Order adopted August 20, 2014: Amendment of the Commission's Rules to Provide Spectrum for the Operation of Medical Body Area Networks, ET Docket No. 08-59, First Report and Order and Further Notice of Proposed Rulemaking, 29 FCC Rcd. 10662 (2014). Available at: https://apps.fcc.gov/edocs_public/attachmatch/FCC-14-124A1_Rcd.pdf

90. Chavez-Santiago, R. *et al.*, Propagation models for IEEE 802.15.6 standardization of implant communication in body area networks, in: *IEEE Communications Magazine*, vol. 51, no. 8, pp. 80–87, August 2003. http://wcsp.eng.usf.edu/papers/UWBBasics.doc.

91. Barroca, N., Borges, L., Velez, F., Goncalves, V., Balasingham, I., Cognitive radio for medical body area networks using ultra wideband. *IEEE Wirel. Commun.*, 19, 4, 74–81, 2012.

92. Nepa, P. and Rogier, H., Wearable Antennas for Off-Body Radio Links at VHF and UHF Bands: Challenges, the state of the art, and future trends below 1 GHz. *IEEE Antennas Propag. Mag.*, 57, 5, 30–52, Oct. 2015, doi: 10.1109/MAP.2015.2472374.

93. Ullah, S., Higgins, H., Braem, B., Latre, B. *et al.*, A comprehensive survey of wireless body area networks. *J. Med. Syst.*, 36, 1065–1094, 2012.

94. Latre, B., Braem, B., Moerman, I., Blondia, C., Demeester, P., A survey on wireless body area networks. *Wirel. Netw.*, 17, 1–18, Jan. 2011.

95. IEEE 802.15 Working Group Document, IEEE 802.15-08-0407-02, Lewis, D., 802.15.6 Call for applications - Response Summary, July 2008.

96. de Vicq, N., Robert, F., Penders, J., Gyselinckx, B., Torfs, T., Wireless body area network for sleep staging, in: *IEEE Biomedical Circuits and Systems Conf. (BIOCAS 2007)*, pp. 163–166, 2007.

97. Reinvuo, T., Hannula, M., Sorvoja, H., Alasaarela, E., Myllyla, R., Measurementof Respiratory Rate with HighResolution Accelerometer and Emfit Pressure Sensor, in: *Proceedings 2006 IEEE Sensors Applications Symposium*, 2006.

98. Ionescu, C.M. and Copot, D., Monitoring Respiratory Impedance by Wearable Sensor Device: Protocol and Methodology. *Biomed. Signal Process. Control.*, 36, 57–62, 2017 Jul. Available from: http:// dx.doi.org/10.1016/j.bspc.2017.03.018.

99. Fang, Y., Jiang, Z., Wang, H.A., Novel Sleep Respiratory Rate Detection Method for Obstructive Sleep Apnea Based on Characteristic Moment Waveform. *J. Healthc. Eng.*, 2018, 1902176, 10, 2018.

100. da Costa, T.D., d. F.F. Vara, M., Cristino, C.S., Zanella, T.Z., Neto, G.N.N., Nohama, P., Breathing monitoring and pattern recognition with wearable sensors, in: *Wearable Devices-the Big Wave of Innovation*, IntechOpen, London, United Kingdom, 2019.

101. Kumar, A., Albreem, M.A., Gupta, M., Alsharif, M.H., Kim, S., Future 5G Network Based Smart Hospitals: Hybrid Detection Technique for Latency Improvement. 8, 153240–153249, 2020. IEEE Access.

102. Kumar, A., Gupta, M., Le, D.N., Aly, A.A., PTS-PAPR Reduction Technique for 5G Advanced Waveforms Using BFO Algorithm. *Intell. Autom. Soft Comput.*, 27, 3, 713–722, 2021.

103. Meena, K., Gupta, M., Kumar, A., Analysis of UWB Indoor and Outdoor Channel Propagation. *2020 IEEE International Women in Engineering (WIE) Conference on Electrical and Computer Engineering (WIECON-ECE)*, IEEE, pp. 352–355, 2020.

104. Gupta, M., Chand, L., Pareek, M., Power preservation in OFDM using selected mapping (SLM). *J. Stat. Manage. Syst.*, 22, 4, 763–771, 2019.

105. Aliverti, A., Wearable technology: role in respiratory health and disease. *Breathe*, 13, e27–e36, 2017.

106. Xiao, Z. *et al.*, An Implantable RFID Sensor Tag toward Continuous Glucose Monitoring. *IEEE J. Biomed. Health Inform.*, 19, 3, 910–919, May 2015, doi: 10.1109/JBHI.2015.2415836.

107. Bahramiabarghouei, H., Porter, E., Santorelli, A., Gosselin, B., Popović, M., Rusch, L.A., Flexible 16 Antenna Array for Microwave Breast Cancer Detection. *IEEE Trans. Biomed. Eng.*, 62, 10, 2516–2525, Oct. 2015, doi: 10.1109/TBME.2015.2434956.

108. Ito, K., *Recent small antennas for medical applications. Int Workshop on IWAT*, vol. 08, pp. 1–4, Chiba, March 2008.

13

Received Power Based Jammer Localization Using Unscented Kalman Filtering

**Waleed Aldosari[1], Muhammad Moinuddin[1,2*],
AbdulahJeza Aljohani[1,2] and Ubaid M. Al-Saggaf[1,2]**

[1]Center of Excellence in Intelligent Engineering Systems (CEIES), King Abdulaziz University, Jeddah, Saudi Arabia
[2]Electrical and Computer Engineering Department, King Abdulaziz University, Jeddah, Saudi Arabia

Abstract

Localizing a jammer in wireless connection networks has become a crucial research problem. The main reason is that launching a jammer and blocking channels between two transceivers has become easier with the advancement of communication technologies. The aim of a jammer is to transmit its signal towards the target channel to decrease the signal noise ratio, which results in degradation in the system bit error rate, and even it can completely block the communication link between the sender and receiver. Thus, it becomes an important task to detect the jammer location that can be helpful later to stop such an attack. This chapter proposed Unscented Kalman Filtering (UKF) based on jamming power received by the node. A thorough simulation was performed to evaluate and compare the performance of UKF to the Extended Kalman Filter (EKF).

Keywords: Localization, jammer, wireless communication, UKF, EKF, detection, JRSS, SNR

Corresponding author: mmsansari@kau.edu.sa

Arun Kumar, Manoj Gupta, Mahmoud A. Albreem, Dac-Binh Ha and Mohit Kumar Sharma (eds.) *Wearable and Neuronic Antennas for Medical and Wireless Applications*, (249–258)© 2022 Scrivener Publishing LLC

13.1 Introduction

Wireless Sensor Networks (WSNs) contains a vast number of distributed nodes. Sensors in the wireless network are self-configurable, randomly deployed, and infrastructure-less. Specific sensors are deployed in a physical environment to observe sounds, pressure, vibration, motion, and target tracking. The component of the sensor is resource-constrained, where the memory is limited, short battery life, short communication range, and limited processing speed. Therefore, sensors are communicating in short distances, and the collected information forwards to the next node until received by the destination or the sink node. Since WSNs are designed to share the wireless medium, it causes serious security challenges. One security threat is the Denial-of-Service attacks (DoS), including jamming attacks and eavesdropping attacks [1, 2].

In a jamming attack, the external node transmits a high power signal to block the communication channel between the sender and receiver. The jamming attack is intended to decrease the network performance by increasing the noise power, which causes a drop in the Signal to Noise Ratio (SNR) at the receiver side. In the communication system, the receiver node has to receive a packet with SNR larger than the system threshold value y, and any packet with SNR lower than y will be rejected and discarded, and a request to retransmit will be sent to the sender. Nodes located inside the jamming region lose their communication link to their neighbors as the jammer keeps transmitting its signal and occupies their channel [3]. Therefore, a transmitting node keeps sensing the channel to transmit its collected data unit, its battery depleted, or the jammer physically removed or turned off. Nodes located outside the jamming region may lose some of their neighbor nodes and still having a communication link to the unaffected nodes. Therefore, we utilize the boundary nodes near the jamming region to capture the jamming signal and track the jammer during a jamming attack [4].

Jamming attacks are classified into two main types: an elementary level and an advanced level. At the elementary level, the jammer is operating and transmitting its signal in the time domain. One type of elementary level is constant jammer. It transmits the jamming signal continually and randomly to make the target channel busy. While the deceptive jammer is transmitting a regular packet towards the target channel, all nodes within the jammer transmitting range switch to receiving mode and process received data until their battery is exhausted. The last type of elementary level or the time domain attack is the random jammer, where the jammer switches between transmit and sleep mode randomly [5]. The advanced level or frequency domain attack includes barrage jammer and spot jammer. Both are operating in the frequency domain where the jammer is sensing the channel to find all available channels then adjusted its power to be higher than the target node. Barrage jammer may target more than one channel simultaneously while the spot targets a single channel during a jamming attack [6].

13.2   Related Work

Jammer localization techniques are divided into two main categories: range-based and range-free. The range-free technique is utilizing the change of network topology and node position to estimate the jammer location. Several algorithms have been proposed using range-free approaches [7–10]. The jammer location accuracy depends on the node density and jammer transmission power. Therefore, a range-free technique causes inaccurate position estimation when few affected nodes or jamming target the network with high power. In this chapter, we are utilizing a range-based method to track a jammer by estimating its received power. Many algorithms used the Jamming Received Signal Strength (JRSS) or jamming information captured by sensors such as Received Signal Strength (RSS), Time of Arrival (ToA), Angle of Arrival (AoA), and Time Difference of Arrival (TDoA) [11].

Many algorithms have been studied to detect the jammer location using the Extended Kalman Filter (EKF), Practical Filter (PF), and Unscented Kalman Filter (UKF). We focused in this chapter on designing a UKF for tracking a jammer using JRSS only in WSN and compare it to an EKF. UKF is a recursive approach using a statistical linearization method and based on unscented transform. Researchers presented many algorithms related to UKF. In [12], UKF based on sensor fusion is given. In the Line of Sight (LoS) environment, the localization accuracy was high compared to the Non-Line-of-Sight propagation (NLoS). The localization accuracy in NLOS was low due to the change of the signal characteristics and a significant amount of noise received by the node. To improve the localization accuracy in NLOS, the author presented a Modified Kalman Filter (MKF) based on UKF. The Linear Kalman filter is utilized to smooth the measurement range and the UKF to track the unknown node.

To improve the localization accuracy, the author in [13] presented a Gaussian sum unscented Kalman filter (GSUKF) and a Gaussian Sum

Particle Filter (GSPF) based on sensor fusion. They utilized the range and angle to track the jammer and used it as an input measurement to the UKF. The Gaussian sum is derived from the sensor model based on the positive and negative information and used to compute the weight directly from Gaussian sum likelihood.

EKF based jammer localization is presented in [14, 15], where the EKF is designed to locate the jammer using JRSS only. The jamming power is estimated from the log-normal shadowing model, where the path loss factor is assumed in a free space environment. Another method presented by the authors is Distance to Signal Noise Ratio (DSNR) based jammer localization. The first step towards detecting the jammer's location is Moore–Penrose Pseudoinverse (MPP). The main concept of MPP is to estimate the jammer location by estimating the jammer's transmission power. Last, an error minimization technique is presented to reduce the error and improve the jammer localization accuracy.

13.3   System Model

In wireless communication, the signal is susceptible to several challenges when propagating between the transmitter and receiver due to obstacles and surrounding objects. The signal is suffering from reflection, diffraction, and scattering, which causes changes in the signal characteristics [16–18]. According to Friis equation, the received power can be expressed in terms of other system parameters as follows:

$$P_r(d) = P_t \frac{G_r G_t \lambda^2}{(4\pi d)^2 L} \tag{13.1}$$

The power received by the receiver node is represented as $P_r(d)$ at distance d. The node's transmission power measured at a reference node depicted as P_t, the receiver and transmitter gains are represented as G_r and G_t respectively. L is the path loss factor, and λ is the wavelength. The log-normal shadowing model is an extension of the Friis equation to measure the power received affected by the surrounding objects and the path loss as:

$$P_r(d) = P_t + K - 10n \log_{10}(d) + X_\sigma \tag{13.2}$$

where K is a constant value and depends on antenna characteristics, the gaussian noise represented as X_σ, and n is the environmental factor, which depends on the physical location where the signal propagates to the receiver [19, 20].

13.3.1   Unscented Kalman Filter (UKF)

The extended Kalman filter is a method for the data integration and evaluation of parameters. It is utilized for a non-linear system by linearization to make it linear by taking the derivative of the non-linear model using the Jacobian matrix. However, in many cases, it is difficult to reach even to the first-order derivative. Therefore, UKF is used to offer better performance compared to EKF. The unscented transformation is the main part of UKF to estimate the statics of the random variables. Namely, to estimate the mean of the random variable, the covariance, the sigma points used, and weights. Therefore, the second-order can be reached without the need to calculate the Jacobian matrix.

In this work, we localize a jammer utilizing JRSS captured by boundary nodes or the tracker. Therefore, the system model is described as:

$$X_k = f(X_{k-1}) + W_k \tag{13.3}$$

where X_k is the state vector at time k and $w_k \sim N(0, Q_k)$ is the process noise. The measurement vector is r epresented as:

$$Z_k = h(X_{k-1}) + V_k \tag{13.4}$$

where the V_k is the measurement noise with covariance $R_k, h(X_{k-1})$ is the non-linear function. The calculation of the sigma points is as follows:

$$x_{k-1}^0 = \hat{x}_{k-1} \tag{13.5}$$

$$x_{k-1}^i = \hat{x}_{k-1} + (\sqrt{(N+\lambda)P_{k-1}})^i \tag{13.6}$$

$$x_{k-1}^i = \hat{x}_{k-1} - (\sqrt{(N+\lambda)P_{k-1}})^i \tag{13.7}$$

where N is the dimension of the state, and λ is the scaling factor and calculated as:

$$\lambda = \alpha^2(N+\kappa) - N \tag{13.8}$$

The α and κ are constants. The weights for sigma points are computed as:

$$w_m^0 = \frac{\lambda}{N+\lambda} \tag{13.9}$$

$$w_c^0 = \frac{\lambda}{N+\lambda} + \left(1 - \alpha^2 + \beta\right) \tag{13.10}$$

$$w_m^i = w_c^i = \frac{\lambda}{2(N+\lambda)} \tag{13.11}$$

Therefore, the UKF for prediction and updating steps become as follows:

$$x_{k-1}^i = F_{k|k-1} x_{k-1}^i + W_k \tag{13.12}$$

$$\hat{x}_{k|k-1} = \sum_{i=0}^{2N} w_m^i x_{k-1}^i \tag{13.13}$$

$$P_{k|k-1} = \sum_{i=0}^{2N} w_c^i \left(x_{k-1}^i - \hat{x}_{k|k-1}\right)\left(x_{k-1}^i - \hat{x}_{k|k-1}\right)^T + Q \tag{13.14}$$

Furthermore, the updating model of UKF can be defined as:

$$y_{k-1}^i = h\left(x_{k-1}^i\right) \tag{13.15}$$

$$\hat{z}_{k|k-1} = \sum_{i=0}^{2N} w_m^i y_{k-1}^i \tag{13.16}$$

$$P_{zz} = \sum_{i=0}^{2N} w_c^i \left(y_{k-1}^i - \hat{z}_{k|k-1}\right)\left(y_{k-1}^i - \hat{z}_{k|k-1}\right)^T + R \tag{13.17}$$

$$P_{xz} = \sum_{i=0}^{2N} w_c^i \left(x_{k-1}^i - \hat{x}_{k|k-1}\right)\left(y_{k-1}^i - \hat{z}_{k|k-1}\right)^T \tag{13.18}$$

$$K_k = P_{xz}\left(P_{zz}\right)^{-1} \tag{13.19}$$

$$\check{x}_k = \check{x}_{k|k-1} + K_k\left(z_k - \check{z}_{k|k-1}\right) \tag{13.20}$$

$$P_k = P_{k|k-1} - K_k P_{zz} (K_k)^T \qquad (13.21)$$

13.4 Simulation and Results

To evaluate the performance of UKF compared to EKF, we deployed four nodes in different locations. The jammer is moving in 2-dimensional space (x,y) with constant acceleration and variant velocity at each time step. The nodes are set to capture the JRSS at each time step, where the steps are equal to 600 steps. The jammer starts at position (0,0) and boundary nodes located at positions (0,5), (3,10), (8,12), and (15,20). The jammer's transmission power is set to −35 dBm, and the environmental factor n equal to 2. UKF parameters are evaluated to get the best estimation compared to EKF. The state vector $[x,y,v_x,v_y]$ and the measurement vector $[P_1,P_2,P_3,P_4]$ where   is the jamming power received by the boundary at time. Figure 13.1 shows the original trajectory compared to UKF and EKF. The red line represents the original trajectory, the blue line the output of UKF, and the pink line is the EKF. Figures 13.2 and 13.3 are illustrations of the position error on x and y respectively.

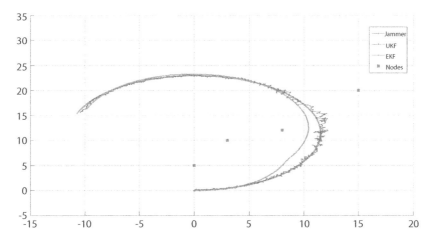

Figure 13.1 Representation of the comparison between the original trajectory, UKF, and EKF.

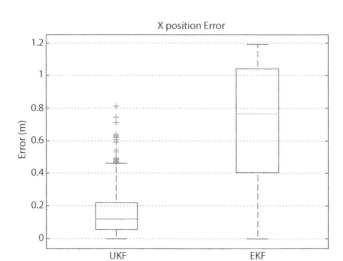

Figure 13.2 X position error.

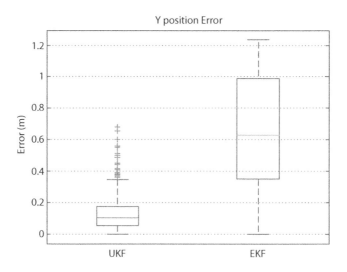

Figure 13.3 Y position error.

13.5 Summary

In this chapter, we have presented UKF based on the jamming power received by different boundary nodes, where the power received follows the log-normal shadowing model, and it is affected by a significant amount of noise. Nodes located close to the jamming region may capture the JRSS and forward it to the base station where the UKF performs the localization algorithm. We compared our work to EKF. The UKF performed better than EKF and can track the jammer accurately.

References

1. Pelechrinis, K., Iliofotou, M., Krishnamurthy, S.V., Denial of service attacks in wireless networks: The case of jammers. *IEEE Commun. Surv.*, *13*, 2, 245–257, 2010.

2. Grover, K., Lim, A., Yang, Q., Jamming and anti–jamming techniques in wireless networks: A survey. *Int. J. Ad Hoc Ubiquitous Comput.* 17, 197–215, 2014.

3. Shiu, Y.S., Chang, S.Y., Wu, H.C., Huang, S.C.H., Chen, H.H., Physical layer security in wireless networks: A tutorial. *IEEE Wirelo. Commun.*, *18*, 2, 201166-74.

4. Xu, W., Ma, K., Trappe, W., Zhang, Y., Jamming sensor networks: attack and defense strategies. *IEEE Netw.*, *20*, 3, 41–47, 2006.

5. Aldosari, W. and Zohdy, M., Tracking a Jammer in Wireless Sensor Networks and Selecting Boundary Nodes by Estimating Signal-to-Noise Ratios and Using an Extended Kalman Filter. *J. Sens. Actuator Netw.*, *7*, 4, 48, 2018.

6. Pang, L., Guo, P., Chen, X., Xue, Z., Tracking The Mobile Jammer Continuously in Time by Using Moving Vector, in: *2017 10th International Symposium on Computational Intelligence and Design (ISCID)*, vol. 1, pp. 43–48, 2017, December.

7. Blumenthal, J., Grossmann, R., Golatowski, F., Timmermann, D., Weighted centroid localization in zigbee-based sensor networks, in: *2007 IEEE international symposium on intelligent signal processing*, pp. 1–6, 2007, October.

8. Wang, T., Wei, X., Sun, Q., Hu, F., GSA-based jammer localization in multi-hop wireless network, in: *2017 IEEE International Conference on Computational Science and Engineering (CSE) and IEEE International Conference on Embedded and Ubiquitous Computing (EUC)*, vol. 1, pp. 410–415, 2017, July.

9. Liu, H., Liu, Z., Chen, Y., Xu, W., Determining the position of a jammer using a virtual-force iterative approach. *Wirel. Netw.*, *17*, 2, 531–547, 2011.

10. Wang, Q., Wei, X., Fan, J., Wang, T., Sun, Q., A step further of PDR-based jammer localization through dynamic power adaptation, 2015.

11. Pelechrinis, K., Koutsopoulos, I., Broustis, I., Krishnamurthy, S.V., Lightweight jammer localization in wireless networks: System design and implementation, in: *GLOBECOM 2009-2009 IEEE Global Telecommunications Conference*, pp. 1–6, 2009, November.

12. Ullah, I., Shen, Y., Su, X., Esposito, C., Choi, C., A localization based on unscented Kalman filter and particle filter localization algorithms. *IEEE Access*, 8, 2233–2246, 2019.

13. Hou, J., Jing, Z.R., Yang, Y., Target tracking in standoff jammer using unscented Kalman filter and particle fiter with negative information. *J. Shanghai Jiaotong Univ. (Science)*, 19, 2, 181–189, 2014.

14. Aldosari, W., Zohdy, M., Olawoyin, R., Tracking the Mobile Jammer in Wireless Sensor Networks Using Extended Kalman Filter, in: *2019 IEEE 10th Annual Ubiquitous Computing, Electronics & Mobile Communication Conference (UEMCON)*, pp. 0207–0212, 2019, October.

15. Kumar, A., Albreem, M.A., Gupta, M., Alsharif, M.H., Kim, S., Future 5G Network Based Smart Hospitals: Hybrid Detection Technique for Latency Improvement. *IEEE Access*, 8, 153240–153249, 2020.

16. Kumar, A., Gupta, M., Le, D.N., Aly, A.A., PTS-PAPR Reduction Technique for 5G Advanced Waveforms Using BFO Algorithm. *Intell. Autom. Soft Comput.*, 27, 3, 713–722, 2021.

17. Meena, K., Gupta, M., Kumar, A., Analysis of UWB Indoor and Outdoor Channel Propagation. *2020 IEEE International Women in Engineering (WIE) Conference on Electrical and Computer Engineering (WIECON-ECE)*, IEEE, pp. 352–355, 2020.

18. Gupta, M., Chand, L., Pareek, M., Power preservation in OFDM using selected mapping (SLM). *J. Stat. Manage. Syst.*, 22, 4, 763–771, 2019.

19. Aldosari, W., Zohdy, M., Olawoyin, R., Jammer Localization Through Smart Estimation of Jammer's Transmission Pwer, in: *2019 IEEE National Aerospace and Electronics Conference (NAECON)*, pp. 430–436, 2019, July.

20. Cheffena, M. and Mohamed, M., Empirical path loss models for wireless sensor network deployment in snowy environments. *IEEE Antennas Wirel. Propag. Lett.*, 16, 2877–2880, 2017.

Index

Printed and bound by CPI Group (UK) Ltd, Croydon, CR0 4YY

27/10/2024

14580131-0001